THE BATTLE OF GETTYSBURG

A BRIEF LOOK AT LEE, LINCOLN, AND THE BLOODY TURNING POINT IN THE CIVIL WAR

ERIC PORTERFIELD

The Battle of Gettysburg

A Brief Look at Lee, Lincoln, and the Bloody Turning Point in the Civil War

Eric Porterfield

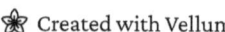 Created with Vellum

CONTENTS

INTRODUCTION: UNFINISHED BUSINESS

"In this enlightened age, there are few I believe, but what will acknowledge, that slavery as an institution, is a moral & political evil in any Country."

— ROBERT E. LEE, DECEMBER 27, 1856, IN A LETTER
TO HIS WIFE, MARY ANNE LEE

When President Abraham Lincoln gave his now-famous address at the Gettysburg battlefield on November 19, 1863, he spoke of this new nation conceived in liberty and equality a mere "four score and seven" years earlier. America's unfinished business – the travesty of slavery – drove Americans to the madness of war. Brother fought against brother, patriot against fellow patriot. Robert E. Lee and Abraham Lincoln were each honorable in their own right. Both men faced difficult choices. What motivated Lee to take such a brash move in marching deeply into the North? And what motivated Lincoln to switch generals before the fighting started in southern Pennsylvania?

CIVIL WAR UNION STRATEGY

The event that triggered the Civil War occurred in the early morning hours of April 12, 1861. The Confederate Army attacked the U.S. military garrison at Fort Sumter, South Carolina. The next day, the U.S. surrendered to the Confederates without a single man killed. While the opening salvos were essentially bloodless, the Civil War stands as the bloodiest war in this nation's history. Four years later – almost to the date – approximately 620,000 Americans had died, more than in both World Wars combined.

Fort Sumter further polarized both North and South. Seven states had already seceded by the time of the Confederate attack on the garrison. Newly-elected President Abraham Lincoln called for 75,000 volunteers to quash the rebellion. Four more states joined the Confederacy.

Though the Civil War formally began in April 1861, the war was not in full swing until spring 1862. The early battles of the war were fought in places like Virginia, Kentucky, Minnesota, Missouri, and Oklahoma. Missouri and Kentucky were two of the southern states that had not seceded. They remained, however, sharply divided. Both states had dual governments, and their citizens fought one another, Missourian against Missourian and Kentuckian against Kentuckian. Most of the battles in Virginia occurred in the region that would later become West Virginia, splitting off from the northernmost of the southern belligerents. Citizens of northwestern Virginia officially declared their desire to remain in the Union on June 20, 1861. The new state of West Virginia was admitted into the Union on June 20, 1863, less than two weeks before Gettysburg.

While the battle of Gettysburg raged, General Ulysses S. Grant laid siege to the Confederate's most significant remaining stronghold at Vicksburg on the Mississippi River. Grant had been assaulting Vicksburg for six solid weeks. One day after the Union won in Gettysburg, Grant won in Vicksburg, setting a turning point in the overall war. Both wins proved to be sober grounds for celebrating America's 87th birthday on July 4, 1863.

The North had devised a simple, four-pronged plan to settle their conflicts with the South:

- Union General McClellan would command the primary effort in Virginia by moving on the state's capital, Richmond.
- Forces in Ohio, one of the three most populous states, would push through Kentucky to take Tennessee.
- The newly formed Department of the Missouri (formed out of the larger Department of the West) would split the Confederacy into two parts by pushing southward along the Mississippi River.
- Union forces in Kansas would attack the Confederacy from the center of the continent.

Lee's surge deep into the North interrupted the Union's careful plan and threatened to upend the North's superiority. Bringing the war directly to the relatively safe Northerners would have created panic at a time when Lincoln's task was already complicated enough.

1

LEE'S CHALLENGE

"I can anticipate no greater calamity for the country than a dissolution of the Union. It would be an accumulation of all the evils we complain of, and I am willing to sacrifice everything but honor for its preservation. I hope, therefore, that all constitutional means will be exhausted before there is a resort to force.

"Secession is nothing but revolution. The framers of our Constitution never exhausted so much labor, wisdom, and forbearance in its formation, and surrounded it with so many guards and securities, if it was intended to be broken by every member of the Confederacy at will. It is intended for 'perpetual Union,' so expressed in the preamble, and for the establishment of a government, not a compact, which can only be dissolved by revolution, or the consent of all the people in convention assembled. It is idle to talk of secession: anarchy would have been established, and not a government, by Washington, Hamilton, Jefferson, Madison, and all the other patriots of the Revolution....

"Still, a Union that can only be maintained by swords and bayonets, and in which strife and civil war are to take the place of brotherly love and kindness, has no charm for me. I shall mourn for my country and for the welfare and progress of mankind. If the Union is dissolved and the Government disrupted, I shall return to my native

state and share the miseries of my people, and, save in defense will draw my sword on none."

— ROBERT E. LEE, JANUARY 23, 1861, IN A LETTER TO
HIS SON, G.W. CUSTIS LEE

Robert E. Lee (1807–1870) did not want secession; neither did he wish for slavery to continue. Why did he become the leader of the South's military in northern Virginia? His past and the environment he grew up in help explain his decisions.

MAN OF CHARACTER

Robert Edward Lee was born on the Stratford Hall Plantation, Westmoreland County, Commonwealth of Virginia, in 1807. The original Lee ancestor moved from Shropshire, England, in 1639. The Lees were one of the oldest families in the state. Robert E. Lee's mother, Anne Hill Carter Lee (1773–1829), also came from a prominent Virginia family. His father, Henry Lee III (1756–1818), had been an officer in the Revolutionary War and had given a eulogy at George Washington's funeral in 1799.

Henry had also served as Virginia's 9th governor (1791–1794) and as a member of Congress representing Virginia (1799–1801). Henry's financial well-being was significantly reduced by the Panic of 1796–1797, fueled by land speculation in the fledgling U.S. Several years later, Henry declared bankruptcy, spending a year in debtor's prison.

In 1812, when war with Britain seemed imminent, Henry was gravely injured in Baltimore, Maryland, while helping a friend ward off an attack by pro-war zealots. Recuperating at home from his injuries proved impossible, so he sailed to the tropical West Indies, hoping the climate would improve his health. Six years later, in 1818, he died in Georgia en route to his Virginia home.

When Henry left home, Robert was only five years old. A relative named William Henry Fitzhugh (1792–1830) allowed the Lee family to stay at his Ravensworth country estate in northern Virginia. Thus, even though his father had been ruined financially and, through ill health, incapable of caring for his family, at least one Virginia relative was sufficiently well off to help. Fitzhugh served several terms in Virginia state government.

In 1825, on Fitzhugh's recommendation, Robert E. Lee entered the academy at West Point with a focus on engineering. He graduated second in his class of 45 cadets and was one of only nine who finished with zero demerits. Class rank and a lack of demerits speak to his self-discipline and intelligence.

Lee served with distinction in the Mexican-American War (1846–1848), earning several promotions throughout the conflict. During the 1850s, Lee became superintendent of the West Point Academy. During his tenure, his eldest son attended and ultimately graduated at the top of his class.

In October 1859, when radical abolitionist John Brown captured the Harper's Ferry, Virginia federal arsenal, President James Buchanan sent Lee with several detachments to quell the violence.

In 1860, when a group of Mexican bandits attacked the military in south Texas, Lieutenant Colonel Robert E. Lee was sent to put down the invaders from across the Rio Grande River. After Lee had finished his assignment, one of the Texas Rangers who had worked with him, John Salmon "Rip" Ford (1815–1897), wrote that Lee was "dignified without hauteur, grand without pride... he evinced an imperturbable self-possession, and a complete control of his passions... possessing the capacity to accomplish great ends and the gift of controlling and leading men."

When war broke out between North and South, Lee felt duty-bound to serve his native state of Virginia and his new, adopted country—the Confederate States of America.

A HARD CHOICE

Robert E. Lee was torn between two loyalties. One loyalty was to the country in whose army he had spent his adult life as an officer. Like all officers, he had sworn an oath to defend its Constitution against all enemies, foreign and domestic. His other loyalty lay with his native Virginia, where his family ties predated the United States by more than a century. Virginia was as much a part of his identity as the Lee family name.

"Mr. Blair, I look upon secession as anarchy. If I owned the four millions of slaves in the South, I would sacrifice them all to the Union; but how can I draw my sword upon Virginia, my native state?"

— ROBERT E. LEE, APRIL 18, 1861, RESPONDING TO AN OFFER FROM PRESIDENT LINCOLN, RELAYED BY FRANCIS PRESTON BLAIR, TO MAKE LEE MAJOR-GENERAL COMMANDING THE DEFENSE OF WASHINGTON DC, AS QUOTED IN *LIFE AND CAMPAIGNS OF GENERAL ROBERT E. LEE* (1866)

Why did Robert E. Lee refuse the President's offer to defend his country? Why did he profess an undying devotion to his "native State"? Before the Revolutionary War, each state was a semi-sovereign nation that jealously guarded its sovereign rights.

"I have fought against the people of the North because I believed they were seeking to wrest from the South its dearest rights. But I have never cherished toward them bitter or vindictive feelings, and have never seen the day when I did not pray for them."

— ROBERT E. LEE, AS QUOTED IN *THE AMERICAN SOUL: AN APPRECIATION OF THE FOUR GREATEST AMERICANS AND THEIR LESSONS FOR PRESENT AMERICANS*, BY CHARLES SHERWOOD FARRISS (1920)

The United States, as a federal collection of states, semi-sovereign, individual units that are bound into a Constitutional compact was understood two centuries ago. A key distinction can be found in the terms national and federal. A well-known law dictionary distinguishes between the two in this way:

"A national government is a government of the people of a single state or nation, united as a community by what is termed the 'social compact,' and possessing **complete and perfect supremacy** over persons and things, so far as they can be made the lawful objects of civil govern-

ment. A federal government is distinguished from a national government by its being the government of *a community of independent and
sovereign states, united by compact.*" [emphasis added]

Before the Civil War, the country commonly referred to as "America"
was frequently called "these" United States. After the Civil War, the
country became increasingly referred to as "the" United States. Here,
the distinction may seem subtle, but it played a part in the motivations
that drove men like Lee to rebellion.

Devotion to one's nearest neighbors is at least as old as any devotion
to a national identity. In the 5th century B.C., in Ancient Greece, the
population of a city like Athens was no more than about 250,000
people. Athens was a major city; today, it would be considered relatively
small, only slightly larger than a town. In that time and place, the city-
state was one's primary focus. A citizen was Athenian, Spartan, or
Corinthian – the city of one's birth. Greeks fiercely defended their cities
against each other but pooled their resources when outside forces
threatened the otherwise squabbling city-states.

In the middle of the 19th century, New York City had a population of
nearly 1,200,000, while Washington, D.C. had only about 75,000. When
the Civil War started, Boston had only about 180,000 people, while
Richmond, Virginia (the state's capital) had only about 38,000 people.
For this country's first 87 years, the state was one's identifying heritage,
while the collection of these United States was still too new for many to
grasp.

Allegiance to one's native state was as important in 19th century
A.D. America as was allegiance to one's city in 5th century B.C. Greece.

Lee's allegiance to his state was a powerful motivator, even in the
face of his professed opposition to slavery.

"So far from engaging in a war to perpetuate slavery, I have rejoiced that slavery is abolished. I believe it will be great for the interests of the South. So fully am I satisfied with this, as regards Virginia especially, that I would cheerfully have lost all I have lost by the war, and have suffered all I have suffered, to have this object attained."

— ROBERT E. LEE, MAY 1, 1870, IN A STATEMENT TO JOHN LEYBURN, AS QUOTED IN *R. E. LEE: A BIOGRAPHY*, BY DOUGLAS SOUTHALL FREEMAN
(1934)

How could a man like Lee — a top graduate and later superintendent of West Point, a U.S. Army officer who repeatedly defended his country — betray the country to which he had sworn a solemn oath?

Once Lee had committed to serve his native state and the new country to which it had allied itself, his destiny was forever tied to the Confederacy and its desperate attempt at survival.

LEE'S STRATEGY

Historians have considered many reasons why Lee carried the war — fought principally to that point on southern soil — to the North. Lee was not the type of man to play defense if he could avoid it. When Lee's government wanted him to come to the defense of the men and fortifications at Vicksburg, Mississippi, Lee refused. Rather than fight a defensive battle against Grant in Vicksburg, he was determined to carry the war to the Union.

He had made bold moves in the North before Gettysburg. In September 1862, Lee attacked the North across the Potomac River into Maryland. The ensuing battle at Antietam Creek, near Sharpsburg, Maryland, was the bloodiest single day of the war. Twenty-three thousand soldiers died in 12 hours of savage combat.

Perhaps Lee hoped to motivate pacifists in the North. By threatening Baltimore, Washington D.C., or even Philadelphia, he might drive the peace-loving Democrats of the North to insist on an end to the war. Lincoln was not the type of leader to bow to public pressure. Lee may have depended too heavily on political pressure than on military strategy.

Lee may also have tired of fighting a defensive war in his home state. He knew that finding provisions for his army was becoming increasingly difficult in the depleted Virginia countryside. Rich farmland in Pennsylvania could provide his troops with an improvement in sustenance.

Perhaps he wanted revenge for his losses in Maryland the previous year during his first foray into northern territory. His losses at Antietam Creek, Maryland, may still have weighed on him.

One motive may have touched a deeper nerve. Robert E. Lee had attached his destiny and honor to his native state of Virginia. Less than two weeks before Gettysburg, a significant portion of his state had joined the Union—his current enemy. With so much of his identity tied to Virginia, a part of him may well have felt an inner division. What might have happened if Virginia had remained whole and rejoined the Union, departing the Confederacy? Would Lee have resigned immediately?

General Lee wrote Confederate President Jefferson Davis five days after he had entered Maryland in 1862—a year before Gettysburg. What he wrote provides insight into his motives and strategy. In his letter, dated September 8, 1862, he stated that "for more than a year both sections of the country have been devastated by hostilities which have brought sorrow and suffering upon thousands of homes, without advancing the objects which our enemies proposed to themselves in beginning the contest."

It had been half a century since the British had assaulted the country, and many remained alive who had suffered from the War of 1812. Lee felt it was already time to offer peace to the North with Confederate independence as its basis. The offer should come from a position of strength, "[m]ade when it is in our power to inflict injury upon our adversary." Lee intended to cause more damage in the North than the enemy did in the South. He told his President that a peace proposal "would show conclusively to the world that our sole object is the establishment of our independence, and the attainment of honorable peace."

Lee argued that if the Lincoln administration were to reject peace, Lincoln's supporters would realize that continuing this costly war fell on the shoulders of the Republicans, not the Confederacy. The general predicted that voters would take their passions to the polls in two months "to determine... whether they will support those who favor a prolongation of the war, or those who wish to bring it to a termination, which can but be productive of good to both parties without affecting the honor of either."

Before Gettysburg, General Lee wrote and expressed his beliefs and strategy.

"If we can baffle them [the North] in their various designs this year & our people are true to our cause... our success will be certain.... [and] next year there will be a great change in public opinion at the North. The Republicans will be destroyed [in the next presidential election] & I think the friends of peace will become so strong as that the next administration will go in on that basis. We have only therefore to resist manfully."

— ROBERT E. LEE, 1863 (SEARS, 2004)

Two other elements help explain Lee's push into the North. One involved Lee's recent successes, while the other involved Confederate desire for foreign recognition.

Before Lee's devastating loss at Antietam in Maryland in 1862, Lee defeated the Union armies at the Second Battle of Bull Run in Virginia in late August. Before advancing into Pennsylvania in 1863, General Lee prevailed in the Battle of Chancellorsville in Virginia. Both times, he attacked the North when his men's morale was high.

In 1863, after two years of fighting, much of it in Virginia, farms in the region had been stripped bare of provisions. Like the British during the Revolutionary War nearly a century earlier, the Confederate army would need to forage to feed the army of 70,000 plus. And like the American patriots had done with the French and Spanish during the Revolutionary War, the Confederacy desperately wanted recognition from European nations in order to negotiate aid. In the war of independence from Britain, without the French, the Americans would likely have struggled valiantly for several more years before losing to the superior British army. The Confederacy was in a similarly tight situation. The North's greater population and more substantial industrial capabilities left the South at a severe disadvantage—a shortcoming President Jefferson Davis and his administration wished to improve with foreign recognition and aid. Whatever other reasons General Lee might have had, a show of superior grit and strength from the increasingly desperate South might have helped to force an early settlement before the South ran out of food, munitions, and men.

WHY NOT WASHINGTON, D.C. OR NEW YORK CITY?

Before the Revolutionary War, military strategy commonly focused on taking the opposition's capital city. The British learned the hard way that Americans did not play by such rules. Even so, the capital remained important. An attack on the capital might have a devastating effect on morale.

Though Union forces made a point to place the main body of their forces between Lee and the District of Columbia, the notion that Lee might attack the capital city itself was a genuine concern for Lincoln and his generals. The capital district was immediately adjacent to the enemy in Virginia.

If Lee had managed to attack Washington and set fire to its buildings, would forces within the government have turned on Lincoln?

New York City had, for a short while in American history, acted as the country's capital city. It had long established itself as the economic center of the young republic. The city also had the largest population of any urban center in America, even at that time.

Perhaps reaching that far into enemy territory would have proven foolhardy. New York City was an additional 180 miles or so East and North of Gettysburg, depending on the route taken. Philadelphia was considerably closer at 110 miles. General Lee's army was already 90 miles from Virginia.

No one knows how much resistance Lee would have experienced if he had left the Union forces behind, chasing after him. And no one knows precisely how much fear he might have generated throughout the North.

THE PATH TO PENNSYLVANIA

With his second invasion of the North in mind, Lee divided his infantry among three of his generals—A.P. Hill (1825–1865), James Longstreet (1821–1904), and Richard S. Ewell (1817–1872). General Jeb Stuart (1833–1864) led the cavalry.

Lee's forces left the vicinity of Fredericksburg, Virginia, on June 3, 1863, heading west from the lower Potomac River toward the Blue Ridge Mountains.

At Brandy Station, the two sides fought a minor battle on June 9.

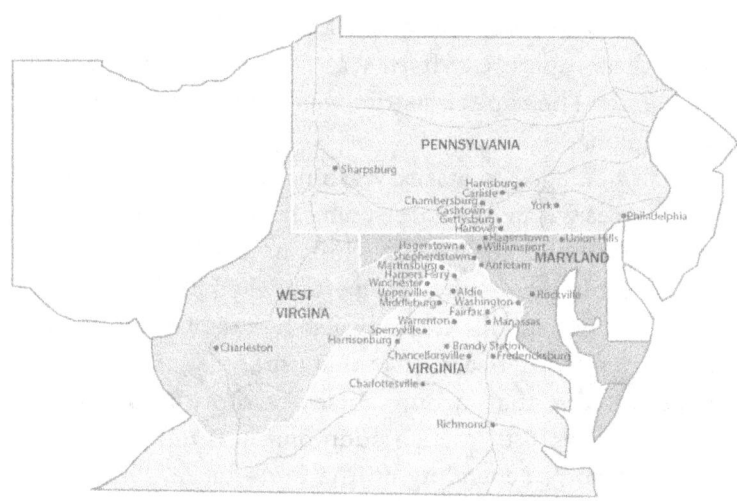

GENERAL A.P. HILL

From Brandy Station, Virginia, General Hill's forces went northward to Upperville, tangling with Union forces on June 21. Then, they pushed northward to cross the Potomac River into Maryland at Shepherdstown, only about ten miles upstream from Harper's Ferry. From there, they went north to Hagerstown, northward into Pennsylvania, and eastward to Gettysburg.

GENERAL JAMES LONGSTREET

General Longstreet moved northward from Brandy Station, crossing the mountains at Snicker's Gap and crossing the Potomac River at Shepherdstown. From there, they traveled northward through Hagerstown, Maryland, and northward, again, to Chambersburg, Pennsylvania. Longstreet then took his men eastward through Gettysburg and York and on to Wrightsville by June 28. From there, they returned to Gettysburg.

GENERAL RICHARD S. EWELL

General Ewell's forces raced past Sperryville, crossing the Blue Ridge Mountains at Chester Gap, then north through the Shenandoah Valley. On June 13–15, they fought at Winchester, Virginia, just south of the West Virginia panhandle. From there, they continued north through Martinsburg, West Virginia, and crossed the Potomac River at Williamsport into Maryland.

Ewell's forces then headed northeast to Hagerstown, Maryland, and then north to Chambersburg, Pennsylvania, by June 28.

From there, Ewell's army followed the Cumberland Valley to Carlisle, Pennsylvania, about 25 miles north of Gettysburg. He then moved on to Harrisburg, the state capital, by June 29, and back to Carlisle by July 1—the first day of the Battle at Gettysburg.

GENERAL J.E.B. STUART (CAVALRY)

General Stuart's cavalry went northward from Brandy Station, Virginia, to Warrenton. From Warrenton, his forces split and continued north along parallel paths, encountering resistance at Aldie on June 17 and Middleburg on June 17–19. From there, Stuart made a brash move. He took his cavalry southward, then eastward around Manassas Junction. He then moved northward past Fairfax to Dranesville to cross the Potomac River, only about 10 miles upriver from Washington, D.C. By this maneuver, he found himself between the Union army and the capital city. Then Stuart went a short distance eastward to Rockville, Maryland, then northward to Westminster, Maryland, Union Hills, and into Pennsylvania to clash with the Union forces at Hanover on June 30. From there, Stuart took his cavalry eastward to York, to the northwest to Carlisle, Pennsylvania, and then southward to Gettysburg.

For over a week, Stuart was out of touch with his commander, jeopardizing the entire mission.

OVERVIEW OF THE PUSH NORTHWARD

General Lee felt obligated to find a quick solution to the months-long conflict. He also felt obligated to do the best possible job for the Confederacy. Though the Davis administration wanted Lee to turn south to help the forces at Vicksburg fend off the assaults by General Ulysses Grant, Lee refused. Perhaps Lee felt he could better aid Vicksburg by drawing forces away to Pennsylvania. Lee could also accomplish several other goals at the same time, including: feeding his hungry troops, giving Virginia a rest, putting the Union on the defensive, and securing the South a big enough win to gain recognition and funding from Europe. If Providence was on their side, his bold move might force the Union to seek an early peace.

When Lee lost touch with his cavalry, led by General Jeb Stuart, he lost a big part of his "advanced warning system," so to speak. Nevertheless, Lee also had spies in nearby Maryland to help feed him much-needed intelligence. After all, Maryland had been a Southern state—one of a few that did not want to leave the Union, though many of its citizens still held Southern sympathies.

In late June, a spy provided Lee much-needed information about the location of the Army of the Potomac. First, he learned that the Union army was no longer led by General Joseph Hooker (1814–1879) but by General George Meade (1815–1872). He also learned that the Union army had taken up a position near Frederick, Maryland.

General Lee had expected the Union to respond to his incursion, and now he had solid intelligence on that response. This intelligence allowed him to plan his next moves without as much worrisome guessing.

Ewell's soldiers had been preparing to cross the Susquehanna River from strategic locations at York and Carlisle. Now, however, Lee ordered him to move in a different direction—Cashtown or Gettysburg.

Longstreet and Hill had been prepared to march northward from Chambersburg and Greenwood, respectively. Now, Lee had them move toward the East to Cashtown.

By concentrating all of his forces east of Pennsylvania's South Mountain, General Lee would have a strategic advantage, allowing him either to defend from a Union attack or to mount an offensive against Meade's forces.

Crossing the Susquehanna River was akin to Julius Caesar crossing the Rubicon. Such a move would put the Confederate forces within relatively easy striking distance of Philadelphia.

Early Monday morning, June 29, 1863, General Meade sent two brigades of cavalry under General John Buford to scout ahead of the Union army. Then, Meade began to move his men northward into Pennsylvania. His forces outnumbered those of Lee by as much as 30,000 men—by some estimates, 104,000 to Lee's 75,000. Union forces included roughly 36,000 cavalry compared to the Confederacy's 9,500.

Besides protecting the country's capital, Meade wanted to provoke Lee into a battle before the southern general could have his forces cross the Susquehanna.

On Tuesday, Union General Buford's cavalry pushed back Confederate General Hill's troops, which were approaching Gettysburg. Even so, Hill ordered General Heth to take his division into the town of Gettysburg on Wednesday.

In the meantime, General Buford kept most of his forces in Gettysburg, awaiting reinforcements from General Meade. Buford had realized the strategic value of the town.

TEST YOUR UNDERSTANDING

Answers to the following questions may be found in the Appendix.

1. Because Lee's father, Henry, left when he was five, who was the father figure in his life?

- *President George Washington*
- *President James Buchanan*
- *William Henry Fitzhugh*
- *John Salmon Ford*

2. Where did Robert E. Lee's greatest loyalty seem to reside?

- *The Constitution of these United States.*
- *The Confederacy.*
- *Virginia.*
- *The institution of slavery.*

3. What military academy did Lee attend, where he later served as superintendent?

- *West Point.*
- *Virginia Military Institute.*
- *Valley Forge Military Academy.*
- *None of the above.*

4. Why would Lee attack so deeply into the North after he had lost so severely the previous year at Antietam?

- *Morale was high after their win at Chancellorsville.*
- *The South needed a significant win to prove to European nations that they had a chance to maintain their independence and to gain not only recognition but also material and financial aid.*

- *Lee's army needed food and supplies, and foraging in southern Pennsylvania could give them that.*
- *All of the above.*

5. How did General Lee move his forces into Pennsylvania?

- *At a Blitzkrieg pace to startle the Union forces.*
- *By dividing his infantry into three groups under generals Ewell, Hill, and Longstreet.*
- *By skirting close to Washington, D.C., to distract Union forces into rushing to defend the capital.*
- *All of the above.*

SUMMING IT UP

General Robert E. Lee had loyally served his country as an Army officer. His ultimate loyalty, however, seems to have laid with his native state of Virginia.

After two years of war, Lee realized the South needed a speedy solution. His men also needed the food and supplies they could no longer find in Virginia. And the Europeans were waiting to see if the Confederacy could prove itself in battle. Would it maintain its independence? General Lee decided he could advance these goals by boldly attacking the North on their territory. He hoped to cross the Susquehanna River and threaten the symbolic heart of the republic, the birthplace of liberty, in 1776.

Of course, President Lincoln did not take kindly to another invasion of the North.

2

LINCOLN'S CHOICE

"[N]o man is good enough to govern another man without that other's consent. I say this is the leading principle, the sheet-anchor of American republicanism. Our Declaration of Independence says: 'We hold these truths to be self-evident: That all men are created equal; that they are endowed by their Creator with certain inalienable rights; that among these are life, liberty and the pursuit of happiness. That to secure these rights, governments are instituted among men, deriving their just powers from the consent of the governed.'

"I have quoted so much at this time merely to show that, according to our ancient faith, the just powers of governments are derived from the consent of the governed. Now the relation of master and slave is pro tanto *a total violation of this principle. The master not only governs the slave without his consent, but he governs him by a set of rules altogether different from those which he prescribes for himself. Allow ALL the governed an equal voice in the government, and that, and that only, is self-government."*

— ABRAHAM LINCOLN, OCTOBER 16, 1854, SPEECH
AT PEORIA, ILLINOIS, IN REPLY TO SENATOR
DOUGLAS

MAN OF THE PEOPLE

Though Abraham Lincoln had been born in Kentucky, he is frequently associated with the state of Illinois. A popular nickname for the state is "Land of Lincoln." The state motto, however, is more revealing of the times when the state was added to the Union—December 3, 1818: "State Sovereignty, National Union." The motto was adopted two months after state admission. War changed a great many things. So embarrassed were the people of Illinois with their motto after the Civil War that they altered the state seal to place the last two words on top.

Though Lincoln grew up in poverty, he had the grit and determination to educate himself, ultimately becoming a lawyer. With a successful practice in Springfield, Illinois, he also became a leader of the more conservative Whig Party, a representative in the Illinois state legislature, and an Illinois representative to the U.S. Congress. For a short while, he returned to his law practice. But in 1854, he became enraged by the passage of the Kansas-Nebraska Act that threatened to spread slavery throughout the American territories. Not long afterward, with the waning influence of the Whig Party, Lincoln joined the new Republican Party, gaining a name for himself in the 1858 debates with Democrat Stephen A. Douglas. Perhaps it was good fortune that Lincoln lost the senate race because he could concentrate his efforts in 1860 on a bid for the presidency.

Lincoln inherited a nation torn by slavery.

"A house divided against itself cannot stand. I believe this government cannot endure, permanently half slave and half free. I do not expect the Union to be dissolved—I do not expect the house to fall— but I do expect it will cease to be divided. It will become all one thing or all the other."

— ABRAHAM LINCOLN, DURING A SPEECH AT THE
SPRINGFIELD, ILLINOIS, REPUBLICAN STATE
CONVENTION, JUNE 16, 1858, WHILE ACCEPTING
THEIR NOMINATION FOR SENATE

The first sentence of this quote is from the Bible, where Jesus Christ is responding to his accusers who thought that his power over devils had come from the Devil himself.

> *"And Jesus knew their thoughts, and said unto them, Every kingdom divided against itself is brought to desolation; and every city or house divided against itself shall not stand."*
>
> — MATTHEW 12:25

DIRE CHOICES

When Abraham Lincoln was elected as the 16th president of these United States, he took office surrounded by a fragile political scene.

Outgoing President James Buchanan had long been a Democrat and globalist who had pushed for American domination through the Manifest Destiny doctrine. Buchanan had been one of three U.S. ambassadors to Europe who drew up the Ostend Manifesto for acquiring Cuba from Spain by force if necessary. This plan created a scandal in Europe at the time.

Lincoln, however, did not possess such grandiose motives. While he did favor keeping the Union intact, he did not start his presidency with the use of force. Instead, he started with calm consultation with the South and compassion for his men stationed there.

Abraham Lincoln had a keen sense of logic and could frequently spot the logical fallacies used by others. In his response to the infamous Dred Scott decision of the federal Supreme Court, he lambasted the chief justice.

> *"We believe...in obedience to, and respect for the judicial department of government. We think its decisions on Constitutional questions, when fully settled, should control, not only the particular cases decided, but the general policy of the country, subject to be disturbed only by amendments of the Constitution as provided in that instrument itself. More than this would be revolution. But we think the Dred Scott decision is erroneous.... If this important decision had been made by the unanimous concurrence of the judges, and without*

any apparent partisan bias, and in accordance with legal public expectation, and with the steady practice of the departments throughout our history, and had been in no part, based on assumed historical facts which are not really true; or, if wanting in some of these, it had been before the court more than once, and had there been affirmed and re-affirmed through a course of years, it then might be, perhaps would be, factious, nay, even revolutionary, to not acquiesce in it as a precedent."

— ABRAHAM LINCOLN, JUNE 26, 1857, SPEECH AT
SPRINGFIELD, ILLINOIS, IN RESPONSE TO THE
DRED SCOTT DECISION

Certainly Lincoln vehemently opposed slavery.

"Slavery is founded in the selfishness of man's nature—opposition to it, in his love of justice. These principles are an eternal antagonism; and when brought into collision so fiercely, as slavery extension brings them, shocks, and throes, and convulsions must ceaselessly follow. Repeal the Missouri Compromise—repeal all compromises— repeal the Declaration of Independence—repeal all past history, you still can not repeal human nature. It still will be the abundance of man's heart, that slavery extension is wrong; and out of the abundance of his heart, his mouth will continue to speak."

— ABRAHAM LINCOLN, OCTOBER 16, 1854, SPEECH
AT PEORIA, ILLINOIS, IN REPLY TO SENATOR
DOUGLAS

Lincoln also idolized Henry Clay, a public servant from Kentucky who had owned slaves his entire life but who had sought to implement a "gradual emancipation."

"Unlike hundreds of other eulogies to Clay, Lincoln's highlighted the Kentuckian's vigorous engagement with slavery throughout his political life. He emphasized that Clay, from the beginning of his public career, 'ever was, on principle and in

feeling, opposed to slavery.' Acknowledging the paradox that Clay was a slave owner, Lincoln declared that he had nonetheless been 'in favor of gradual emancipation of the slaves in Kentucky.' Lincoln admired Clay for opposing 'both extremes' on slavery: those who would 'shiver into fragments the Union' and those who would 'tear to tatters the Constitution' in their desire to overthrow slavery immediately. Lincoln was intent to 'array his name, opinions, and influence,' against 'an increasing number of men' who, Lincoln feared, were beginning to assail 'the declaration that "all men are created free and equal." ' "

— RONALD C. WHITE, JR., *A. LINCOLN: A BIOGRAPHY* (2010)

One popular view of history is that the South was ready to secede if Lincoln became President. However, the true picture of mid-19th-century times is more nuanced. Hostilities broke out between North and South only once Confederate President Jefferson Davis attacked Fort Sumter, South Carolina—a crime by the laws of the land. He did this secretly while Lincoln had been quite open with the South Carolina governor about his intention to resupply his men stationed there without new arms or ammunition.

Although others orchestrated the conditions for the Civil War, it was Lincoln's responsibility to manage the process through to completion. As Commander-in-Chief, it was up to him to supervise the generals and admirals who conducted the war.

JUDGING BY MERIT

Joseph Hooker (1814–1879) had been assigned to command the Army of the Potomac on January 26, 1863. After his loss at the Battle of Chancellorsville (April 30–May 6, 1863), Hooker wanted to attack the Confederate capital of Richmond, Virginia, perhaps wanting to save face with an easy win. Lincoln denied this request. The Old-World strategy of ending a war by taking the opponent's capital had never worked in the Americas. Hooker was not thinking strategically. Lee could do far more significant damage by assaulting Philadelphia or Baltimore than Hooker could accomplish with Richmond.

Instead, Lincoln wanted Hooker to chase Lee's army and defeat the Confederate general before he could set off the political powder keg which the North had become. The Union president had already put down riots and dangerous rhetoric in northern cities. If Lee could burn Philadelphia, he might create a level of panic in the North that Lincoln would be unable to manage.

Hooker seemed reluctant to chase after Lee. After his loss at Chancellorsville, Hooker did not have the confidence he had once portrayed. He requested that troops stationed in northern Virginia be transferred to the Army of the Potomac to reinforce his efforts against Lee. In late June 1863, when Lincoln refused this request, Hooker resigned his command. Lincoln gladly accepted.

Lincoln had not particularly liked Hooker but needed competent men to get the job done. Earlier in the war, Hooker had been quoted by a *New York Times* reporter as saying that he welcomed a dictator and that nothing would get done until we had one. Later, Lincoln wrote the general and told him somewhat unkindly that Hooker had been selected for his position despite his scandalous words.

However, Lincoln was even more disappointed that Hooker was reluctant to chase Lee and finish the job he had failed to complete at Chancellorsville.

Lincoln chose General George G. Meade (1815–1872) to replace Hooker. Though not Lincoln's first choice, on June 28, 1863, Lincoln sent word to Meade of his appointment. Though several other officers under Hooker had outranked Meade, they recommended the junior commander for the job, letting those in charge know that they were willing to serve under Meade despite their superior ranks.

TEST YOUR UNDERSTANDING

Answers to the following questions may be found in the Appendix.

1. What kind of childhood did Lincoln have?

- *He had a middle-class upbringing.*
- *He lived in poverty, having to teach himself.*

- *His parents were wealthy landowners.*
- *He came from a wealthy family which had fallen on hard times.*

2. What was Lincoln's political background?

- *He was a Social Democrat.*
- *Like George Washington, he did not believe in political parties.*
- *He was a religious conservative.*
- *He was a Communist in his youth and a conservative in early adulthood.*

3. What dire choices did Lincoln have to make?

- *Whether or not to challenge the belligerent Confederates for their attack on Fort Sumter, South Carolina.*
- *Whether or not to go to war over the growing hostilities in the South along its border with the North.*
- *Whether or not to force the South back into the Union.*
- *All of the above.*

4. What was Lincoln's view of slavery?

- *That it was okay if restricted to the slave states.*
- *That it was an evil institution based on selfishness.*
- *That the institution should be allowed to die out naturally.*
- *All of the above.*

5. Why did General Hooker resign as commander of the Army of the Potomac?

- *He was emotionally challenged after losing the Battle at Chancellorsville and reluctant to face General Lee again.*
- *He may have been angry that Lincoln did not let him score an easy win by taking the Confederate capital of Richmond, Virginia.*
- *Lincoln disapproved of the requested reinforcements for his army.*
- *All of the above.*

SUMMING IT UP

Abraham Lincoln was a simple man who spoke with profound logic. When he spoke, he was open and authentic, yet he was a capable statesman who only said what needed to be said, guided by wisdom.

He was passionate about the evils of slavery but preferred to defuse that emotional powder keg with time and finesse. However, when provoked, Lincoln was not timid about using decisive, swift force to quell unruly mobs, treasonous publishers, and violence in the face of negotiations. Lincoln was also decisive about his commanding generals, replacing Hooker with Meade when the former was reluctant to chase Lee after the Union loss at Chancellorsville. Meade and Lee would meet in Gettysburg, Pennsylvania just one month later.

3
THE BATTLE BEGINS

"On the morning of July 1 the officers of the regiment cast about rather anxiously for the wherewithal to 'piece out' a breakfast, having lost touch, for two or three days with the wagons of the commissary department. Captain Sigler and a lieutenant or two formally asked permission to send out to buy a sheep, but as there was a prospect of an early movement, this could not be accorded, and they had to be satisfied with the regulation coffee and hard-tack. The field and staff fared no better."

— LT. COLONEL THOMAS CHAMBERLIN, 1895,
*HISTORY OF THE ONE HUNDRED AND FIFTIETH
REGIMENT, PENNSYLVANIA VOLUNTEERS, SECOND
REGIMENT, BUCKTAIL BRIGADE*

General Lee's forces were scattered across Pennsylvania. They were strewn from Wrightsville and near Harrisburg (the state capital) on the Susquehanna River, through Carlisle (30 miles north of Gettysburg) to Chambersburg (28 miles west-northwest of Gettysburg).

Upon learning on June 29 that the Army of the Potomac had left Virginia in pursuit of the Confederate army, Lee ordered all of his forces to assemble at Cashtown, Pennsylvania, some 8 miles west of Gettysburg. At that point, Second Corps, under General Ewell, had already left Chambersburg for Carlisle to the north, arriving there on the first day of battle.

Fighting on the first day—July 1, 1863—started as isolated skirmishes. Troops and equipment continued to arrive throughout much of the day. The great irony involves General Lee's orders—that no one was to engage the Union army until all of the Confederate forces had arrived. However, Third Corps' division commander, General Henry Heth (1825–1899), supposedly sent his junior officer, Brigadier General James J. Pettigrew, in search of supplies, including shoes for his soldiers. Pettigrew encountered Union forces just outside Gettysburg.

Union Brigadier General John Buford (1826–1863), commander of Cavalry Corps' First Division, recognized, early on, that the hills south of Gettysburg—Cemetery Hill, Cemetery Ridge, and Culp's Hill—might offer a strategic advantage to whoever held them. To keep the Confederate forces from acquiring that high ground, he took his dismounted cavalrymen to the ridges west of Gettysburg to delay the approach of the Southern army.

Major General John F. Reynolds (1820–1863), one of the most respected commanders in the Union army, moved his I Corps into position to block the Confederate advance on Gettysburg, assisting Buford's efforts west of the town. Major General Oliver O. Howard (1830–1909) was to follow close behind with his XI Corps of primarily German immigrants.

The first ridge the Confederate Army encountered was Herr's Ridge.

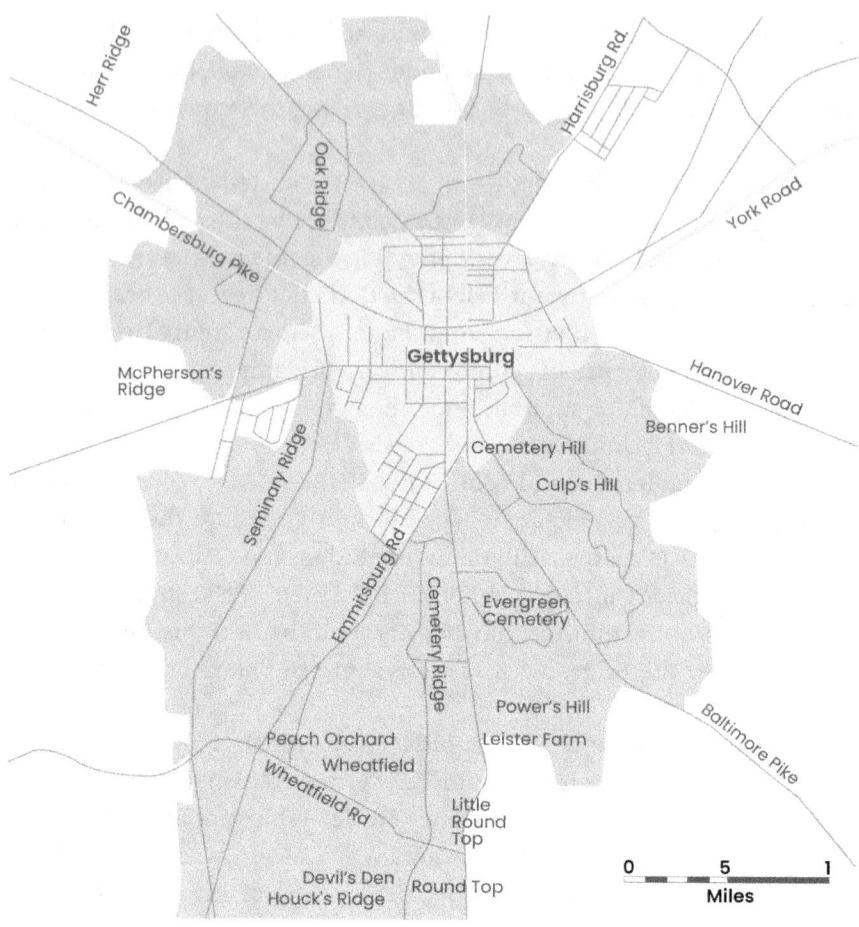

Gettysburg, Pennsylvania

HERR'S RIDGE

On July 1, 1863, the first day of the famous battle, North and South met at Herr's Ridge. Herr's Ridge was the westernmost of three key ridges alongside Chambersburg Pike leading into Gettysburg from the West. Union Cavalry General John Buford needed only to slow down the Confederate advance.

General Buford had positioned his dismounted troops north and west of Gettysburg on the previous evening, knowing full well that his relatively small band of men could not hold off an entire enemy army.

They would, however, guard the roads leading into town from those directions and would block immediate access to the crucial high ground south of town.

Two divisions of the Confederate Third Corps, under General A.P. Hill, would have preferred to march along Chambersburg Pike into town. Buford's men forced the Confederates off the road and into offensive lines of battle, as the Southern soldiers slogged uphill, trying to dislodge the smaller Union forces from their superior, high-ground positions.

Though General Buford had chosen McPherson's Ridge to be his main point from which to disrupt the advancing Confederates, his troops engaged with the enemy before they could get there, commencing the fight at Herr's Ridge. It was the farthest of the three ridges from Gettysburg—roughly two miles from the center of town. Two Confederate brigades, plus a contingent of artillery, clashed with Union forces at Herr's Ridge. Brigadier Generals James Archer and Joseph Davis commanded the brigades.

The Confederate artillery played an important part in dislodging the Union forces. As the artillery moved forward, infantrymen helped to protect them from Union fire. And finally, Buford's men were routed. By 9 a.m., the overwhelming Confederate forces had taken Herr's Ridge.

Buford's mission was accomplished nonetheless; the Confederates were slowed sufficiently for the Union I Corps to reinforce the Union efforts west of Gettysburg.

MCPHERSON'S RIDGE

Later that morning, after Union General Buford's men abandoned Herr's Ridge, they fell back to McPherson's Ridge to keep up their resistance to the Confederate advance. McPherson's Ridge was the second of three key ridges alongside Chambersburg Pike leading into Gettysburg from the West. General Buford needed to continue slowing down the Confederate advance.

While Confederate General Davis pushed toward Gettysburg on the north side of Chambersburg Pike, General Archer advanced on the south side of the road.

At about 9:30 a.m., the Union I Corps, under General Reynolds, arrived from the south onto the battlefield to reinforce Buford's efforts. Buford's 2nd Brigade, under Col. Thomas Devin, kept up their barrage of fire north of Davis's position.

While Davis's brigade advanced onto the ridge in the bottom half of the 10 a.m. hour, the presence of I Corps on top of McPherson's Ridge stalled Archer's advance. Shortly after 10 a.m., General Reynolds was killed. Major General Abner Doubleday took command of the corps.

I Corps' Second Brigade of First Division, under Brigadier General Lysander Cutler, was forced back by Davis's attack and took a position at the bottom of the ridge after 11 a.m.

Davis's advance forced Union Captain James A. Hall, I Corps' Artillery Brigade, to retreat with his six guns. Meanwhile, Union Second Lieutenant John W. Roder advanced with only one gun to cover Hall's retreat by lobbing fire into an unfinished railroad cut, resulting in the capture of 200 Confederate prisoners.

About 11 a.m., Union troops moved along the ridge, just north of Chambersburg Pike, clashing with Davis's brigade, forcing them to retreat.

After 11 a.m., Union troops south of the turnpike pushed westward, down the ridge, forcing the Confederate Army to retreat. Union Brigadier General Solomon Meredith's Iron Brigade (I Corps, First Division, 1st Brigade) captured several hundred Confederate soldiers, including General Archer.

Archer was the first Confederate general of Lee's Army of Northern Virginia to be captured. He had suffered from a lingering illness that had begun during the summer of 1862 and the Confederate invasion of Maryland. Though Archer would survive his year as a prisoner of war, his illness finally took him on October 26, 1864.

Buford was again successful in delaying the Confederates with his dismounted cavalrymen and Union I Corps. Though Union General Reynolds was killed, Confederate General Archer was captured.

SEMINARY RIDGE

On the afternoon of July 1, 1863, Confederate forces pushed the Union Army back through the town and into the hills south of Gettysburg. But the Union Army had delayed Confederate forces long enough so that the Union Army could take the superior, high ground south of Gettysburg.

Seminary Ridge, the third of three critical ridges alongside Chambersburg Pike leading into Gettysburg from the West, was closest to the town. The afternoon of the first day of battle saw heavy fighting between the two armies. The Union forces planned to secure the more defensible high ground at Cemetery Hill, Cemetery Ridge, and Culp's Hill.

Union General Winfield Scott Hancock (1824–1886), nicknamed "Hancock the Superb," had made a name for himself numerous times on the battlefield. At Chancellorsville, his division had protected General Hooker's retreat, and he was wounded for a second time that year. When his corps commander transferred to a different posting in protest over Hooker's behavior, Hancock assumed command of II Corps of the Army of the Potomac.

When the new commander of the Army of the Potomac, General Meade, heard of the death of his friend, General Reynolds, earlier in the day, he sent Hancock to take temporary command of the Union forces at Gettysburg.

When Hancock arrived, General Howard of the XI Corps protested because Hancock was not the most senior officer on the field. Howard had assumed command of all Union forces upon the death of Reynolds. Ironically, this same description ("not the most senior officer") fit General Meade, who would give orders directly after his arrival in the early hours of July 2.

Meade had given Hancock full authority to assess the situation at Gettysburg and to withdraw all forces if necessary. However, instead of withdrawing, Hancock decided to protect the hills south of Gettysburg for their strategic advantage. And this is what Buford's actions earlier in the day had made possible.

After a midday lull in fighting, hostilities picked up again at about 2 p.m., especially after the arrival of Confederate General Early from the northeast of Gettysburg. Confederate General Ewell had also recently arrived from the north.

Union General Carl Schurz had been sent to take Oak Hill, northwest of Gettysburg, but it was too late. Confederate Second Corps' division commander, Major General Robert E. Rodes, had already taken the hill.

When Ewell, the Confederate corps commander, arrived, both he and Rodes surveyed the field. After Carl Schurz failed to take Oak Hill, Ewell and Rodes decided that Union troop movements justified ignoring Lee's orders not to engage with the enemy. Many had already ignored Lee's standing order.

Confederate forces attacked Union emplacements from the west, north, and northeast. Union forces, overextended at Barlow's Knoll were vulnerable from multiple directions at this "salient"—a protrusion in a line of defense that becomes vulnerable to attack from more than one direction. As a result, Union forces were quickly overrun.

General Lee arrived on the field at about 2:30 to witness Rodes fully engaged with the enemy and lifted his earlier restrictions against engagement. Then, General Heth continued his attacks on the Union I Corps from the west, forcing them back to Seminary Ridge.

All across the line, by 4 p.m., Confederate forces pushed the Union army back, through the town and into the hills to the south of Gettysburg. Union forces were losing ground, and commanders ordered all remaining men to retreat to the relative safety of Cemetery Hill.

General Lee recognized the importance of the high ground south of town and supposedly had told General Ewell to take the hill occupied by Union forces "if practicable." Scholars note that there are no known written orders from July 1863 on this matter, and Lee only added the words "if practicable" to his revised report in January 1864.

General Buford, supplemented by I Corps and XI Corps, had delayed Confederate forces long enough so that the Union Army could take the superior, high ground south of Gettysburg.

When the day was done, about 22,000 Union troops had faced off with roughly 27,000 Confederate soldiers. The Union lost nearly 9,000 men, while the Confederacy lost a little more than 6,000. That first day alone, 15,000 Americans had lost their lives.

TEST YOUR UNDERSTANDING

Answers to the following questions may be found in the Appendix.

1. Where would the battle have likely been held if all of Lee's generals had followed orders?

- *Gettysburg.*
- *Cashtown.*
- *York.*
- *Carlisle.*

2. Which Confederate officer was responsible for starting the Battle of Gettysburg?

- *General A.P. Hill*
- *General Henry Heth*
- *General James Pettigrew*
- *All of the above*

3. What critical instructions did Lee give his commanders, along with the order to gather at a specific town?

- *To forage for food along the way.*
- *To harass northerners so they would be more inclined to demand that Lincoln end the war.*
- *Only to engage with the enemy once all Confederate forces were consolidated.*
- *To take the hill occupied by the enemy at Gettysburg, if practicable.*

4. Why did General Buford array his dismounted cavalry west and north of Gettysburg?

- *To spot Confederate forces approaching by one of the main roads.*
- *To disrupt the movements of the Confederate army toward Gettysburg.*
- *To protect the approach to the hills south of Gettysburg, for their strategic importance, until reinforcements could arrive.*
- *All of the above.*

5. Who was the highest-ranking officer to become a prisoner of war on the first day of the battle at Gettysburg?

- *General Robert E. Lee*
- *Lt. General Ambrose P. Hill*
- *Major General Henry Heth*
- *Brigadier General James Archer*

SUMMING IT UP

The battle at Gettysburg began with disobedience of Lee's order to engage with the enemy only after all Confederate forces had been assembled at one location. Initially, that location was Cashtown, about 8 miles west-northwest of Gettysburg.

The alert initiative of General John Buford prepared the way for a Union victory at Gettysburg, delaying the Confederates long enough for the Union to take the high ground. Perhaps General Pettigrew should have withdrawn his divisions at the first sign of Union resistance. After all, he was supposedly sent to Gettysburg for supplies, and Lee's standing order not to engage had yet to be lifted.

There were many instances of courage and good work on both sides of the conflict. General Hancock's taking the high ground south of Gettysburg was strategically sound. Because General Meade had trusted Hancock, even to the point of ordering a complete withdrawal, the battle of Gettysburg could proceed.

Fateful disobedience had forced Lee's hand, and he tried to make the most of it.

4

LEE'S PUSH TOWARD VICTORY

"It must be remembered that we make war only upon armed man, and we can not take vengeance for the wrongs our people have suffered without... offending against Him to whom vengeance belongeth...."

— ROBERT E. LEE, FROM PUBLISHED ORDERS,
QUOTED IN *THE DEATH OF A NATION*, BY
CLIFFORD DOWDEY, 1958

From what we know of General Robert E. Lee's behavior, he seems to have been a man of honor who also valued the honor of others. Lee amended his report months after Gettysburg, which helped General Ewell to save face after failing to take the high ground south of town.

We also see this in Lee's somewhat forgiving reaction when he learned that his orders not to engage had been disobeyed. Instead of firing his generals, he merely lifted his earlier restrictions. Perhaps it was more practical to have quiet words of rebuke in private and to help keep morale high with the successes of the first day of fighting. After all, they had the Yankees on the run.

THE PLAN

Lee's plan was simple: carry the momentum of the South's wins at Chancellorsville weeks earlier and Gettysburg the day before to crush the Union forces. Morale was high, and Lee was anxious to score a significant victory in Union territory to force an early end to the war. Lee's plan, however, was ill-fated because his assessment of the Union placements did not match reality.

When all the combatants had assembled on the battlefield, there were nearly 94,000 Union troops against fewer than 72,000 Confederate soldiers. The North had a clear advantage.

General Longstreet suggested that the army should move to a position blocking Meade's communication and supply lines to Washington, forcing the Union general to attack on more favorable ground. But Lee did not want to sour the morale of his troops by giving up ground for which they had shed blood.

By the morning of July 2, the Union forces had become well entrenched in their high-ground defenses. Whatever advantage the Army of Northern Virginia had possessed was lost the previous day when General Ewell had not found it "practicable" to take the high ground before the Union military could fortify its positions.

Union forces were concentrated in a tight knot with a semi-circular defensive front, arcing from Little Round Top, through Cemetery Ridge, across Cemetery Hill, and onto Culp's Hill. Union command, supply, and reinforcement lines were thus short and fast.

The Confederate forces were spread out, encircling the Union positions, making communications, supply, and reinforcements more difficult.

The specific details of Lee's plan were based on faulty intelligence because his cavalry—a key reconnaissance element—was absent. General Jeb Stuart had taken an unplanned detour through Rockville, Maryland, to pick up supplies. Though the supplies were helpful, the critical function of gathering accurate information was far more vital. When Stuart finally showed up in Gettysburg in the early afternoon, Lee, no doubt, had some choice words for his wayward cavalryman.

General Lee gave the following instructions:

General Longstreet, commander of First Corps, was to start the attack by sending two of his three divisions at the Union's left flank, approaching them at an angle by moving along Emmitsburg Road. Each division would remain perpendicular to the road, with General McLaws (1821–1897) on the west and General Hood (1831–1879) on the east. With this oblique attack, Lee hoped to collapse the enemy's left flank, allowing the Confederate forces to take Cemetery Hill.

The Third Corps division under General Anderson would then attack at the heart of the Union line where the enemy held Cemetery Ridge.

The problem with Lee's plan stemmed from the location of Union forces. They were not arrayed along one side of Emmitsburg Road but on the hills and ridge farther east.

General Lee's morning surprise attack suffered from several delays. General Hood was missing a brigade and had to wait until it arrived. And when that brigade finally made it to Gettysburg, it had to move through an indirect path to avoid detection by Union lookouts.

Finally, by 4 p.m., the soldiers were all in their positions, ready to begin.

ANOTHER WRINKLE IN LEE'S PLAN

In the meantime, unbeknownst to General Lee, one of Union General Meade's officers had disobeyed orders.

Meade had ordered General Daniel Sickles (1819–1914), commander of III Corps, to take up a position on Cemetery Ridge, next to General Hancock and his men, anchoring his left flank on Little Round Top, at the southernmost end of the Union defensive arc.

Nearly a mile in front of III Corps stood ground slightly higher than that Sickles and his men occupied. No doubt, he feared a repeat of past mistakes at Chancellorsville. There, Sickles relinquished higher ground that Lee's army later used as a platform to launch deadly artillery rounds. The higher ground, here, was the Peach Orchard. Without permission or communication, General Sickles moved his men to what he thought was a better, more strategic location. He moved his men to not only the Peach Orchard but also extended his line back through the Wheatfield and across to Devil's Den—all well in front of Cemetery Ridge and Little Round Top, where he should have been.

This move by Sickles would end his military career. Not only was he disobeying orders, but he was also not telling command of the change, thus jeopardizing the entire battle. Sickles had a history of impulsive behavior, not the least of which was murdering a man on the streets of Washington, D.C., for allegedly having an affair with his wife. That homicide took place only four years earlier—February 27, 1859.

The worst part of Sickels's mistake at Gettysburg involved the strategic blunder of creating a salient in a defensive line, making the line vulnerable to attack from multiple directions.

When General Meade discovered that Sickles and his men were out of position, he let the III Corps' commander understand his mistake. But Meade told Sickles to stay because it was too dangerous to retreat to Cemetery Ridge with the Confederates already in motion.

HOOD ATTACKS

Everything changed when Lee's army discovered that the Union III Corps occupied the Peach Orchard adjacent to Emmitsburg Road. Longstreet started with a 36-gun barrage that lasted half an hour. The plan, then, was to attack the Union army more squarely but *en echelon*— a leap-frogging technique where one regiment would push until its momentum faltered, and the next would move in to keep up the drive against the enemy. Once the shelling had stopped, Hood attacked, telling his men, "Fix bayonets, my brave Texans! Forward and take those heights!" Regrettably, he had not specified which "heights" they were to take.

Not long afterward, Hood was struck by shrapnel and taken out of action. His second-in-command, Brigadier General Evander M. Law (1836–1920), did not know until much later that the division was without a leader. Rough terrain and lack of leadership left regiments moving more or less haphazardly in the general direction of the Union lines.

Five regiments moved on the Round Tops while the remaining four attacked Devil's Den. While Hood's division advanced, Longstreet held back McLaw's division for the next wave of attacks.

DEVIL'S DEN

Devil's Den consisted of a modest rise punctuated with massive boulders. And because General Sickles had moved his troops after noon, his men needed more time to fortify their positions. When the Confederate regiments met Ward's brigade in Devil's Den, both sides were relatively unprotected, and fierce fighting ensued.

Union Brigadier General J. H. Hobart Ward (1823–1903) tried his best to defend the left end of his line from the fierce onslaught, even pulling regiments from the right to protect his beleaguered left.

One immediate objective of the Confederate advance was to destroy the Union artillery on Houck's Ridge, overlooking Devil's Den. In defense, one regimental commander guarding the guns took his men down the ridge, sending the advancing Rebels running. But the Confederates rallied with such a barrage that two-thirds of the Union assault never made it back up the ridge to their starting point.

Well after the battle had begun, some of Confederate General Hood's men saw that a significant gap existed between Ward's brigade and that of Régis de Trobriand (1816–1897). The Confederate brigades of George Anderson (1824–1901) and Henry Benning (1814–1875) slammed into that gap, attempting to gain the upper hand, but the Union resistance proved too formidable. Anderson was wounded and had to be carried to safety by his retreating men.

Some of Benning's men continued down Plum Run Valley, attempting to outflank Ward at the other end of his line of defense. Union General Birney scrambled to send reinforcements to head off the Confederate maneuver. The resulting confrontation became a mutual slaughter. The small valley became known as "Bloody Run" or "Valley of Death."

One Union regiment after another had to fall back. In one instance, as the 40th New York retreated, the 6th New Jersey protected their retreat but lost a third of their men.

Eventually, Union General Ward had to abandon his position. Confederate General Hood's men claimed Devil's Den and the southernmost part of Houck's Ridge.

MCLAWS AT THE WHEATFIELD

Five hundred yards north of Devil's Den and a similar distance north-west of Little Round Top stood the Wheatfield. It lay southeast of the Peach Orchard.

About 5 p.m., thirty minutes after Hood's attack had begun, Longstreet directed General McLaws to send in his first two brigades—that of General William Barksdale (1821–1863), left front, facing the Peach Orchard and that of General Joseph B. Kershaw (1822–1894), right front, facing various fields and hills leading to the Wheatfield.

Before reaching the Wheatfield, Kershaw's brigade encountered Union V Corps, First Division under Brigadier General James Barnes (1806–1869), having only recently arrived to reinforce the battered III Corps, First Division, 3rd Brigade. Stony Hill, near the Rose family farm-house, had been supported by two of Barnes's Union brigades, commanded by Colonel William S. Tilton (1828–1889) and Colonel Jacob B. Sweitzer (1821–1888). Despite the fierce attack, the Union forces held. But then, in a bizarre tactic, Barnes suddenly moved his men back several hundred yards to a point near the Wheatfield Road, facing the Wheatfield, which stood south of the road. The Wheatfield Road was the east-west thoroughfare that ran roughly perpendicular to Emmitsburg Road at the Peach Orchard, past the Wheatfield, and alongside the northern foot of the Little Round Top.

When Kershaw's men reached the Wheatfield, General Anderson's men of Hood's division were already engaged with Union forces there—an overflow from the Confederate attack on the adjacent, north end of Houck's Ridge.

Earlier, when Union General Meade had realized the blunder of III Corps' commander, General Sickles, he also sent one of Hancock's II Corps' divisions—First Division, headed by Brigadier General John C. Caldwell (1833–1912). Caldwell's immediate objectives were to clear the Wheatfield and to drive the Confederates from Stony Hill. He had his 1st Brigade, led by Colonel Edward E. Cross (1832–1863), take care of the Wheatfield, while his 2nd Brigade, led by Colonel Patrick Kelly (c.1822–1864) and his 3rd Brigade, led by Brigadier General Samuel K. Zook (1821–1863), took care of Stony Hill. In the meantime, Caldwell held his 4th Brigade in reserve, a unit led by Colonel John R. Brooke (1838–1926).

When the Union forces clearing the Wheatfield were running out of ammunition, Caldwell activated his 4th Brigade to reinforce their efforts. At about that same time, Union defenses in the Peach Orchard failed and had to withdraw. Confederate General William T. Wofford (1824–1884) led his brigade along Wheatfield Road, taking Stony Hill and then outflanking Caldwell's men in the Wheatfield. In this round of fighting, three generals were mortally wounded—both Cross and Zook under Union commander Caldwell and General Paul Jones Semmes (1815–1863) under Confederate commander McLaws.

In response to Wofford's assault on the Wheatfield, Union General Barnes sent in Sweitzer's brigade. After fierce, hand-to-hand combat, Union forces reclaimed the Wheatfield. Most of V Corps' Second Division, under General Romeyn B. Ayres (1825–1888), arrived to reinforce Sweitzer's possession of the Wheatfield. The win proved temporary, however, for the Confederate forces rallied and sent the Second Division scurrying back to the comparative security of Little Round Top. Throughout their effort to help, the Second Division of V Corps lost an incredible 32% of their 2,613 soldiers from direct fighting in the Wheatfield and while crossing the Valley of Death under the deadly gaze of Confederate sharpshooters stationed in Devil's Den.

The Confederate brigades of Anderson, Kershaw, and Semmes drove the Union Second Division of army regulars from the Wheatfield, past Houck's Ridge, toward Little Round Top. At about 7:30 p.m., Confederate forces had advanced to the Valley of Death and continued to the base of Little Round Top.

Proud Pennsylvanians of V Corps, Third Division, under General Samuel W. Crawford (1829–1892), forced the Confederates back past the Wheatfield and onto Stony Hill. Then, realizing that his forces had extended too far beyond the current Union line, Crawford pulled his men back to the far side of the Wheatfield, effectively leaving the field as a "no-man's land" between enemies.

More than 20,000 men had been engaged in the taking and retaking of the Wheatfield with severe losses in the range of 30%. Though not as infamous as Antietam (September 17, 1862) or the far earlier battle in Texas at the Alamo (1836), the Wheatfield at Gettysburg marks one of the bloodier struggles of modern warfare history—a tragedy within the greater tragedy that was Gettysburg.

CONFRONTATION AT THE PEACH ORCHARD

At the start of Confederate General Kershaw's push to reinforce Hood's Division, Kershaw's left flank turned to protect their advance from Union fire within the Peach Orchard.

The Peach Orchard, along with the corner of Emmitsburg Road and Wheatfield Road, belonged to Union forces under the III Corps' First Division and the 1st Brigade troops under Brigadier General Charles K. Graham (1824–1889). Graham's men were supported by Union artillery, including III Corps Artillery Brigade Lt. Bucklyn, plus Artillery Reserve units under Captain Ames, Captain Hart, and Captain Thompson.

On a day full of ironies, some unknown person shouted the wrong command. Kershaw's left flank suddenly turned eastward, making themselves vulnerable to Graham's men and their supporting cannon fire.

While Kershaw's men attempted to skirt past the Peach Orchard to assist Hood's men, McLaws sent his left flank directly into the Peach Orchard. Hood's men had engaged with the enemy over thirty minutes earlier along Houck's Ridge and Devil's Den. McLaws's left flank included the leading brigade, under Brigadier General William Barksdale, and the trailing brigade, under General William Wofford.

The Peach Orchard was the tip of Union General Sickles's spear aimed at the heart of the Confederate forces. But unlike a fast-moving spear, this salient in the Union line merely sat there, receiving abuse from multiple angles. To be sure, the Union forces were not quiet in their suffering, but Sickles's blunder had made thousands of Union soldiers vulnerable.

Union General Andrew A. Humphreys (1810–1883), commander of III Corps, Second Division, had his men arrayed along the eastern side of Emmitsburg Road from the Peach Orchard, past Trostle Lane, for about a third of a mile. But when Kershaw took his Confederates south of Wheatfield Road, many of them turned their attention and their aim at that enemy movement.

When Barksdale's forces slammed into the Union line, the Yankees were unprepared. The Confederate brigade took advantage of the momentary confusion, wheeling to the left to outflank Humphreys and his men, collapsing their line in minutes.

Union General Graham's men who, minutes earlier, had been thrashing Kershaw's men, were now so overwhelmed that they attempted to fall back toward Cemetery Ridge. Graham did not make it.

While Barksdale chased after the fleeing Union forces, Wofford's men focused on the enemy still defending the Peach Orchard.

As Union troops fled past Trostle Lane, III Corps headquarters at the Trostle barn started to evacuate. General Sickles did not move fast enough and suffered from the heavy fire aimed at the retreating III Corps. He was caught in the right leg by a cannonball and had to be carried out on a stretcher. General Birney assumed command of the rapidly disappearing III Corps.

At about 6 p.m., Confederate Major General Richard H. Anderson (1821–1879), one of Lt. General A.P. Hill's division commanders, continued the *en echelon* attack farther north on Emmitsburg Road, making Humphreys' withdrawal that much more difficult.

Generals Meade and Hancock saw all of this from their positions of relative safety on Cemetery Ridge. They attempted to find reinforcements to help defend their collapsing line, but they had used almost every unit they could spare.

TEST YOUR UNDERSTANDING

Answers to the following questions may be found in the Appendix.

1. What did General Lee do that helped weaken his chances of success?

- *He refused to listen to General Longstreet's suggestion that they draw the Union forces onto more favorable ground.*
- *He had not been more forceful in demanding that General Ewell attempt to take the Union high ground before they could strengthen their defenses.*
- *He relied on scanty intelligence in the absence of his cavalry.*
- *All of the above.*

2. What advantages did Meade's army have over the Confederates?

- *Their lines of supply, reinforcements, and communication were shorter.*
- *The Union army had over 20,000 more men than did the Confederates.*
- *The North held the high ground, which is almost always a military advantage.*
- *All of the above.*

3. Which Union officer disobeyed General Meade's orders?

- *General Hood.*
- *General Hancock.*
- *General Sickles.*
- *General Hooker.*

4. What technique did Confederate General Longstreet use to soften up the Union forces in Union General Sickles's salient?

- *He had General Hood tell his men to fix bayonets before charging.*
- *He took the high ground before the fight to have the superior position.*
- *He fired 36 cannon for thirty minutes.*
- *All of the above.*

5. What bizarre tactic did Union General Barnes perform while defending the Wheatfield in support of General Sickles's III Corps?

- *He attacked the advancing Confederates en echelon.*
- *He came down from the safety of his high-ground defenses to engage the enemy hand-to-hand.*
- *He withdrew his forces from the Wheatfield.*
- *He attempted to outflank the advancing Confederates but failed.*

SUMMING IT UP

The Confederate army's second day at Gettysburg was a costly gamble with nearly as many losses as gains. General Lee needed better intelligence, but his generals did a respectable job adjusting to the ever-changing conditions of the battlefield.

The major blunder of General Sickles and the bizarre move of General Barnes were two of the mistakes which were costly for the Union forces.

Though the Confederacy gained Devil's Den and advanced their line east of Emmitsburg Road past the Peach Orchard, north and south of Wheatfield Road, they lost thousands of men to acquire those precious few yards.

One of the more desperate struggles of the second day involved the coveted high ground known as Little Round Top.

5
LITTLE ROUND TOP

"THE stars were shining brightly on the morning of the 2d when I reported at General Lee's head-quarters and asked for orders. After a time Generals McLaws and Hood, with their staffs, rode up, and at sunrise their commands filed off the road to the right and rested. The Washington Artillery was with them, and about nine o'clock, after an all-night march, Alexander's batteries were up as far as Willoughby's Run, where he parked and fed, and rode to head-quarters to report."

— LT. GENERAL JAMES LONGSTREET, 1896, *FROM MANASSAS TO APPOMATTOX: MEMOIRS OF THE CIVIL WAR IN AMERICA*, CHAPTER XXVII, P.362

The fighting at Little Round Top on the second day of the Battle of Gettysburg is closely tied to the conflict at Devil's Den. However, Little Round Top was even more strategically important.

CONFEDERATE STRUGGLE TO TAKE LITTLE ROUND TOP

Because General Sickles, commander of III Corps, had disobeyed General Meade's orders, the subordinate's men no longer defended Cemetery Ridge and Little Round Top. Rather, they had moved several hundred yards west of that position, anchoring their salient in the Union defensive line at Devil's Den, west of Little Round Top, across Plum Run (Valley of Death).

After General Meade had lambasted Sickles for defying direct orders, he sent Brigadier General Gouverneur K. Warren (1830–1882), his chief engineer, to investigate the situation at the Union army's far left flank. When General Warren arrived at Little Round Top, he noticed Confederate bayonets reflecting sunlight in the distance, southwest of the hill. Realizing that the Confederates would soon be able to swarm the Union flank, he sent several staff officers to round up as many troops as possible to defend the hill.

Union V Corps, General George Sykes sent a messenger to order his First Division commander, General Barnes, to secure the hill. But before the message reached Barnes, it was intercepted by his 3rd Brigade commander, Colonel Strong Vincent (1837–1863). Not waiting for Barnes to act, Vincent ordered his four regiments—20th Maine, 16th Michigan, 44th New York, and 83rd Pennsylvania—to fulfill those orders. Why Vincent did not wait for orders from his commander, we do not know. Considering General Barnes's behavior in abandoning the Wheatfield, perhaps it was best that Vincent did not get permission.

Vincent and his bugler, Oliver Norton, led on horseback, riding ahead of their men to look for the best and safest path to their destination.

Vincent's 3rd Brigade arrived at the summit of Little Round Top only 10 minutes ahead of the Confederate forces. His brigade was now the far-left flank of the entire Army of the Potomac, and he had put the 20th Maine at his far left. He instructed the leader of those 385 men, Colonel Joshua Lawrence Chamberlain (1828–1914), that, no matter what, they were to hold their position.

Keen to the dangerous nature of his position in the line, Chamberlain sent several men, including some sharpshooters, to a location behind a stone wall 150 yards east of the Little Round Top summit.

Meanwhile, Meade's chief engineer, General Warren, continued to recruit regiments to help reinforce the Union left flank.

Brigadier General Evander M. Law of Hood's Division ordered five regiments to take Little Round Top—3 Alabama regiments (4th, 15th, and 47th) and 2 Texas regiments (4th and 5th). That day, these men had already marched some 20 miles in the summer heat. Currently, their canteens were empty, and they needed to refill them, but the orders to move out prevented that.

Confederate 4th and 5th Texas, plus the 4th Alabama regiments, attempted twice to storm the hill but were summarily repulsed each time.

The 15th Alabama, under Colonel William C. Oates (1835–1910), attacked the 20th Maine several times. Each time, the Union soldiers drove them back.

During the battle, Confederate General Law discovered that his commander had been severely injured and that he was now the head of the division. Accordingly, Law sent word to Colonel James L. Sheffield that he was now head of the brigade.

During their last attempt, the 15th Alabama attempted to skirt around the end of the Union position.

Colonel Chamberlain realized that his men, now out of ammunition, would likely be unable to repulse another Confederate attack. He ordered his men to fix bayonets and swarm the enemy position. The plan of his attack started by having his left hold back and then advance to catch up. When the left was in line with the others, the rest of the regiment would move forward, combining a flanking maneuver and a frontal assault. The 20th Maine took many of the 15th Alabama prisoners.

As the rest of the Confederate forces retreated, they suffered at the hands of the previously concealed sharpshooters and other Union riflemen behind the stone wall.

Arriving late in the fighting were the 5th United States Battery D of the Artillery Brigade, under Lt. Charles E. Hazlett (1838–1863), with six guns, and 140th New York. Soon afterward, the 91st and 155th Pennsylvania and the 146th New York arrived to reinforce the Union's southernmost flank.

After the fighting had stopped for the evening, soldiers built up stone breastworks as defensive barriers for the assaults they were certain would come the following day.

TEST YOUR UNDERSTANDING

Answers to the following questions may be found in the Appendix.

1. Who discovered that Little Round Top was unprotected?

- *General Ulysses S. Grant.*
- *General Warren, chief engineer of the Army of the Potomac.*
- *General Sickles.*
- *Colonel Chamberlain, 20th Maine Regiment.*

2. Who took the initiative to secure Little Round Top without waiting for orders or permission?

- *Colonel Strong Vincent, commander V Corps, First Division, 3rd Brigade.*
- *General George Sykes, commander Union V Corps.*
- *General James Barnes, commander V Corps, First Division.*
- *None of the above.*

3. What disadvantages did the Confederate army face in attempting to take Little Round Top?

- *The men were out of water.*
- *Taking the high ground is almost always more difficult than fighting on level ground.*
- *The men had been marching as much as 20 miles that day.*
- *All of the above.*

4. What did General Warren do after Colonel Vincent took the initiative in taking Little Round Top?

- *He took a much-needed break far from the front line.*
- *He continued to find reinforcements for the Union left flank on Little Round Top.*

- *He sent word to General Lee that the Union forces wanted to talk peace.*
- *None of the above.*

5. What was Colonel Chamberlain's last-ditch effort to protect the Union's left flank?

- *He had his men roll boulders down the hill since they were out of ammunition.*
- *He signaled to his sharpshooters to take out the Confederate lead as they moved up the hill.*
- *He asked Colonel Vincent for more ammunition.*
- *He had his men fix bayonets and charge down the hill at the enemy.*

SUMMING IT UP

Little Round Top was a significant point of vulnerability after General Sickles abandoned it. Once General Meade's chief engineer discovered that the hill had been left undefended, he ordered nearby commanders to send whatever men they could to secure it.

Confederate General Law sent several regiments to take Little Round Top, but they failed to capture the hill despite repeated attempts.

By 6 p.m., additional reinforcements had arrived, and after the day's fighting ended, the soldiers started constructing defensive barriers.

Attention was focused on one location on the battlefield more than any other: Cemetery Hill.

6

CEMETERY HILL

"As soon as I saw my men entering the town, I rode forward into it myself, having sent to repeat the order to Smith to advance, and when I had ascertained the condition of things, I rode to the right of it to find either General Ewell, General Rodes, or General Hill, for the purpose of urging an immediate advance upon the enemy, before he could recover from his evident dismay and confusion. Rodes' troops were then entering the town on the right and all plains on that flank had been cleared of the enemy. The enemy, however, held the houses in the edge of the town on the slope of Cemetery Hill with sharpshooters, from which they were pointing an annoying fire into Hays' left, and along the streets running towards the hill."

— LT. GENERAL JUBAL A. EARLY, 1912, *LIEUTENANT GENERAL JUBAL ANDERSON EARLY C.S.A.: AUTOBIOGRAPHICAL SKETCH AND NARRATIVE OF THE WAR BETWEEN THE STATES. WITH NOTES BY R.H. EARLY*, CHAPTER XXIV, P.269

A s one of the high-ground locations closest to the town of Gettysburg, Cemetery Hill remained a point of focus throughout the three-day battle, if not by direct military action, at least by its nature as a definitive milestone. If Lee had taken and secured that hill, the battle of Gettysburg might have been his.

CEMETERY HILL: DAY 1

Brigadier General John Buford had taken some dismounted cavalrymen to the ridges west of town to keep the Confederates from taking the high ground south of town. The commander of the XI Corps, Major General Oliver O. Howard, left both artillery and infantry at Cemetery Hill to secure that valuable high ground to back up Buford's efforts.

On the first day of conflict, the local Union forces were north and west of the town, and more would arrive in the coming hours. The primary fallback position would be Cemetery Hill because of its location and strategic importance.

When the approaching enemy forces proved to be too great, both the I Corps and XI Corps had to move back to the high ground they had secured.

Of all the mistakes made by the Confederacy at Gettysburg, the failure of Lt. General Richard Ewell to take Cemetery Hill that first evening was arguably the greatest. As commander of Lee's Second Corps, Ewell took the blame. Brigadier General William "Extra Billy" Smith (1797–1887), commanding one of the three main divisions under General Ewell and Major General Jubal Early, mistakenly thought Union forces were rapidly approaching from the east. Any Union movements east of their position proved to be minor if they existed at all. Based on Smith's bad information, General Early delayed attacking Cemetery Hill because he did not want his men to be outflanked by an incoming regiment or brigade.

Smith's superiors had already become wary of his performance on the battlefield. Smith was a politician rather than a tactician.

Smith had been governor of Virginia (1846–1849) and a member of Congress representing Virginia (1841–1843 and 1853–1861). His last term in Congress ended just before hostilities broke out between the North and the South.

When Virginia left the Union, Smith had been offered a commission as a general. He had declined the offer, citing his lack of knowledge and experience. But later, when the war came to his doorstep, Smith assisted by taking command of a company when its captain had been killed. Because he had enjoyed this taste of command, he requested a place in the Confederate army and was given the rank of colonel. At that time, he was the oldest officer in the entire Confederate army at 64.

After the battle at Gettysburg, General Early commended all his brigadier generals except Smith. A week later, Smith resigned.

The Confederacy may have won at Gettysburg if General Ewell and his division commander, General Jubal Early, had taken Cemetery Hill before it could have been fortified. Was Smith merely a scapegoat for Ewell and Early's failure? We may never know for sure.

CEMETERY HILL: DAY 2

The second day of fighting began late in the afternoon from what seemed to be unavoidable delays. Though hostilities began toward the Union's left flank—Devil's Den, Peach Orchard, Wheatfield, and Little Round Top—Confederate General Lee wanted both ends of the Union line assaulted.

Leading the Confederate Second Corps, General Ewell was to launch a near-simultaneous assault on the north end of the Union line. Though this was intended to be a minor attack, primarily to distract Union commanders while General Longstreet launched his efforts at the south end of the line, General Lee instructed Ewell to take strategic advantage of any successes. The commander of the Second Corps was to use his discretion in escalating his efforts.

About 4 p.m., when General Ewell heard guns echoing from the south, he began a barrage of artillery fire about a mile northeast of Cemetery Hill, atop Benner's Hill. For three hours, that long-range assault was the totality of Ewell's efforts. Because Benner's Hill is more than 50 feet shorter than Cemetery Hill, the Union artillery stationed on that higher ground had a significant advantage. Though the Union gunners suffered some damage, their retaliation drove Ewell's batteries from their position.

By 7 p.m., the Confederate efforts on the Union left flank were waning in strength; General Ewell decided to implement an attack with his infantry.

According to modern records, sunset on July 2, in Gettysburg, Pennsylvania, occurs at 7:42 p.m. (subtracting an hour for the modern invention of daylight savings), and twilight ends at 8:13 p.m. So, as Ewell began his main infantry effort, he had a little over an hour of light available to him and his men.

First, Confederate General Ewell ordered three brigades to cross Rock Creek and advance up the eastern side of Culp's Hill. As that action got underway, shortly after sunset, he sent another two brigades to approach from the east and to attack the smaller mound, East Cemetery Hill. Meanwhile, he told Major General Robert Rodes (1829–1864) to stand ready to assist with an attack on the larger rise, Cemetery Hill, from the northwest.

Major General Edward "Allegheny" Johnson (1816–1873) moved his three brigades against the Union XII Corps brigade of Brigadier General George S. Greene (1801–1899), defending Culp's Hill from behind a line of breastworks. Though they fought for hours into the darkness, Johnson's men were only able to establish a meager foothold within some rifle pits that Union soldiers had abandoned.

The two brigades sent against East Cemetery Hill were from General Early's division, totaling about 2,100 men. The two commanders were Brigadier General Harry T. Hayes (1820–1876), leading the "Louisiana Tigers," and Colonel Isaac E. Avery (1828–1863), leading Hoke's Brigade (General Hoke had been severely wounded earlier in the year and had been at home recuperating). A third brigade commanded by Brigadier General John B. Gordon (1832–1904) stood ready to assist but ultimately did not engage with the enemy.

Two brigades of Union XI Corps defended East Cemetery Hill. Both were part of Barlow's division, which had come under the command of Brigadier General Adelbert Ames (1835–1933) on the first day at Gettysburg. General Barlow (1834–1896) had disobeyed orders, leaving the position on the line assigned to him to gain higher ground. The enemy had taken advantage of this mistake, routing his division. In the panic of the Union withdrawal, Barlow's men left him for dead. The East Cemetery Hill Union brigades were led by Colonel Andrew L. Harris (1835–1915) and Colonel Leopold von Gilsa (1824–1870). Together, they had a total of 1,150 men.

Before the Confederate assault, Ames had moved one regiment from the left to the center of the line, leaving a gap that the South quickly exploited. But there were other weak spots on the line. The onslaught breached several others. Not long afterward, Confederate soldiers had reached the top of the hill and were fighting in the growing darkness, in hand-to-hand combat with the gunners stationed there.

One Confederate officer would later write that they had "cleared the heights and silenced the guns," but while Union accounts acknowledge the breach of position, they point out that it was short-lived. Cemetery Hill was considered the "keystone of the Union line," and its loss to the enemy "would have certainly required Meade to at least abandon his position" on the field at Gettysburg. Indeed, Cemetery Hill was only about a third of a mile north-northeast of Meade's headquarters at Leister farmhouse. Because the South had failed to secure the hill, Union guns stationed there would, on the following day, help repel the main Confederate attack, known later as "Pickett's Charge."

When Union Generals Schurz and Howard heard the fighting nearby, they moved two regiments to help reclaim the hill and secure the battery. But this move left General Howard's line a little thin, so he requested reinforcements. Union General Hancock ordered one of his brigades from Cemetery Ridge. Just as the Confederate push was weakening, Union reinforcements arrived to sweep the enemy from the hill. As the two regiments chased after the enemy at the base of the hill, they "flopped down" to allow the guns above to send the Rebels a barrage as a parting gift.

The lack of coordination between Confederate divisions made a successful attack virtually impossible. For instance, Rodes's division should have attacked long before Early's assault faltered, but it had not been ready. And when Rodes finally moved on Cemetery Hill, the utter darkness made any effort extremely risky. Ahead, his division faced not only two lines of enemy soldiers protected by stone walls but a formidable battery of Union guns threatened the approach.

Later, General Rodes wrote in his report on the battle that he had not received the needed cooperation from the adjacent division. This may have been due, in part, because the division's commander, Major General William Dorsey Pender (1834–1863), had been severely wounded earlier in the day. But when General Ewell sent someone to ask Pender's replacement, Brigadier General James H. Lane (1833–1907), about his lack of action, Lane replied that he had been ordered to take aggressive action against the enemy if a "favorable opportunity presented" itself to him and his men. Yet, when General Ewell sent word back to Lane that the attack was beginning and thus requested his assistance, Ewell never received a reply to acknowledge the request.

CEMETERY HILL: DAY 3

Throughout the third day of the Gettysburg battle, there were no other Confederate infantry assaults on Cemetery Hill. The primary focus of Confederate efforts was at Cemetery Ridge and Culp's Hill. However, the battery of guns atop Cemetery Hill added to the barrages of canon fire against the attacking Southerners.

General Lee had intended to collapse the Union line northward along Cemetery Ridge, but that plan had proven more difficult than initially anticipated.

AFTER THE BATTLE

For many weeks after the battle, the Pennsylvania state militia camped on East Cemetery Hill. From their tented campsite, they kept up a military presence to discourage curiosity seekers and looters and to secure the battlefield until they could collect all of the weaponry strewn across the field of death. Several hospitals continued to operate after the battle, caring for the wounded, and the militiamen supported the doctors and nurses in their mission.

While the keeper of Evergreen Cemetery was away at war, his pregnant wife, Elizabeth Thorn, a few hired hands, plus her parents took part in digging a total of 105 graves for those killed on or close to Cemetery Hill. This was a small part of the estimated 8,000 deaths of approximately 51,000 casualties during the three-day conflict.

TEST YOUR UNDERSTANDING

Answers to the following questions may be found in the Appendix.

1. Though General Ewell was ultimately responsible for not taking Cemetery Hill on day one, whose incorrect report led to not following through on taking the hill?

- *General Jubal Early.*
- *Colonel Leopold von Gilsa.*
- *General James H. Lane.*
- *General William Smith.*

2. Concerning the officer whose report led to the failure to take Cemetery Hill on the first day, what was his background?

- *He had been a school teacher.*
- *He was a politician.*
- *He had been an orphan with no known kin.*
- *All of the above.*

3. From your answer to question #1 above, what personal deficits may have led to his blunder?

- *Age, because he was the oldest Confederate officer to hold a field command.*
- *Lack of experience in military affairs.*
- *Lack of knowledge of military intelligence and strategy.*
- *All of the above.*

4. What caused General Ewell's near-simultaneous assault on the Union's right to be cut short?

- *The hill from which his artillery barrage was launched was far shorter than Cemetery Hill, giving the Union artillery a compelling advantage in return fire.*
- *His battery was overrun by Union cavalry.*

- *The battery ran out of gunpowder early in the assault.*
- *Heavy rain made it nearly impossible to continue firing cannon.*

5. What severe disadvantage plagued General Rodes in his attempt to take Cemetery Hill?

- *Lack of support from adjacent brigades.*
- *Darkness.*
- *Bad timing because General Early's efforts had already lost their momentum.*
- *All of the above.*

SUMMING IT UP

In battle, the high ground is greatly prized by seasoned military officers. Greater height gives a military force an advantage over their adversaries, all other things being equal. On the first day, General Howard was wise to leave some of his men and cannon on top of Cemetery Hill. Union forces stationed there were fortunate that the Confederate mistake gave them more time to secure their position. Thus, the Confederacy lost their first opportunity to take Cemetery Hill.

Cemetery Hill was the most prominent high ground toward the center of the Union line and stood not far from Union General Meade's headquarters.

The second day of fighting had a very late start. By the time General Ewell committed his infantry to the assault on the Union's right, the sun was already minutes from setting. Poorly coordinated efforts on the part of the Confederate command gave the Union another much-needed advantage, which allowed them to fend off the assaults, even after the crest of the hill had been overrun.

Finally, the second day ended with the Confederate failure to keep the hill once they had reached the summit. They would not get a third chance.

On the third day, Union batteries on Cemetery Hill pounded the Confederate forces on the fields below. After the battle, Cemetery Hill served as headquarters for the clean-up effort by the Pennsylvania militia.

Throughout history, humans have had infantry. Cavalry as a weapon of warfare, however, reached its pinnacle during the Civil War.

7
IMPORTANCE OF THE CAVALRY

"Thirty-year-old J. E. B. Stuart, a cadet at West Point when Lee was superintendent there, had served the Army of Northern Virginia as no other cavalry leader had served an army during the war. Skillful, meticulous, and aggressive in reconnaissance, Stuart, Lee said, had never brought him a piece of wrong information, and his screening of the army was flawless. Perhaps Stuart liked fighting for its own sake a little too much, and a tendency to vainglory led him occasionally into rather gaudy exploits, but he was a dedicated Confederate and an instinctive soldier, and he knew his role in this desperate invasion."

— CLIFFORD DOWDEY, 1958, *LEE AND HIS MEN AT GETTYSBURG: THE DEATH OF A NATION,* CHAPTER 1

A man on horseback has several advantages over an infantryman, including mobility, range of view, and, to a degree, many of the advantages of higher ground.

Cavalry was a largely ignored warfare element between the Revolutionary War and the Civil War. By the end of the Civil War, however, it became a multifaceted, finely-honed military tool.

When the Confederacy fired on Fort Sumter, South Carolina, the Union army had only five cavalry regiments. By the end of 1861, that number had increased to 50.

Cavalries had been used frequently in Europe, and the North and South would expand on the traditional uses of cavalry. Such uses included:

- Reconnaissance—Scouting out the locations of the enemy and maintaining a watchful eye on their movements.
- Screening—Protecting the movements of your infantry, armaments, and supplies from discovery by the enemy.
- Covering—Protecting the most vulnerable parts of your army (both flank and rear), especially the rear, when withdrawing.
- Uncovering—Threatening the flanks and rear of your enemy.
- Shock Charges—Heavy, rapid attacks meant to break the enemy's resolve with the desired result of a rout.
- Raiding—Attacks behind enemy lines using the element of surprise.
- Escorting—Accompanying senior officers to protect the chain of command.
- Delivering—Carrying important messages between command and the lines.

By the end of the Civil War, the cavalry had become a fully-developed fighting force known for its effectiveness and flexibility. As we have already seen, General John Buford had taken his dismounted cavalrymen to a ridge west of Gettysburg to slow down the approaching Confederates. We also saw how the late arrival of J.E.B. Stuart's cavalry had sorely disappointed General Lee, for he needed their reconnaissance capabilities to help him in planning the upcoming battle.

Though the Civil War cavalry would use nearly all of the traditional functions from earlier wars, both in America and in Europe, they would discover new methods to maximize this unique resource.

Near the start of the war, some in command were reticent to make much use of cavalry, for one, because such units were so expensive.

Equipping a cavalry regiment cost around $300,000, and maintenance was also costly, adding $100,000 or more yearly. Adjusted for inflation, the maintenance costs alone would exceed $3 million per year in 2023 dollars.

Northerners, for the most part, had to be trained not only how to ride a horse but also how to behave as an army unit. Cavalrymen in the South, for the most part, had grown up riding horses. Thus, for the first year, the South had an advantage over the North in utilizing cavalry either as a part of a larger infantry unit or as a separate unit. That advantage would evaporate during the war, mainly because the South did not have the resources to supply replacement horses sufficient to cover the needs of a protracted conflict. The North, though its resources were not limitless, could keep sending horses to the front lines throughout the war.

By the battle at Gettysburg, the Union cavalry had become a formidable force with expanded responsibilities. Yet, because there had not been a strong tradition for cavalry in these United States before the Civil War, Congress was reluctant to commit funds for heavy cavalry—the type used in charges. Instead, they preferred light cavalry.

CAVALRY UNITS

During the Civil War, there were four key types of cavalry units.

- Regular Cavalry—Soldiers and officers who fought primarily from horseback using sabers and pistols.
- Dragoons—Soldiers and officers with a mix of fighting styles, both on horseback and dismounted, using sabers, pistols, and carbines.
- Mounted Infantry—Infantrymen and officers who traveled by horse but dismounted to engage the enemy. Their primary weapons were pistols, rifles, and bayonets.
- Irregular Cavalry—Unconventional soldiers and officers, including guerrillas, partisans, and armed civilians who moved and fought on horseback. Though there was no standard weaponry, double-barrel shotguns were popular.

THE GROWING NEED FOR CAVALRY

As the heated rhetoric between the North and South increased in the months leading up to Fort Sumter, both sides had vague notions about what their cavalry should look like. As the war started, both sides may have had little more than mounted infantry. Though it had advantages over the traditional infantry, the cavalry's potential was greatly under-utilized. That would change under the pressures and necessities of war.

Though the governments of the North and the South did not want to commit the resources for heavy cavalry, traditionally required for the powerful charges used by the likes of Napoleon, the use of saber charges in combination with soldiers on the ground wielding rapid-fire carbines, proved to be an effective strategy on the battlefield.

When the Civil War broke out, Brevet Lieutenant General Winfield Scott (1786–1866) had been General-in-Chief of the Union army for 20 years. He had several reasons for dismissing the need for cavalry, including:

- The war would be short, and it would take too long to train a cavalry unit.
- Cavalry units are expensive to form and maintain.
- The government needed equipment stocks for cavalry, so it had nothing to give cavalry units even if they were to be formed.
- His main concern was Virginia, which lacked the proper terrain for European-style tactics with cavalry—thick with woods, uneven ground, and too many fences.

The massive Union loss at the First Battle of Bull Run (July 21, 1861) forced President Abraham Lincoln to disagree with General Scott and order the creation of several more cavalry units. On November 1, 1861, General Scott retired and was replaced as General-in-Chief by Major General George B. McClellan (1826–1885), who held the position for less than six months.

Historians have noted how cavalry techniques developed throughout the Civil War led to many successes. One of the Union Army's best cavalry commanders was General Philip Sheridan (1831–

1888). British Lieutenant General Sir Henry Marshman Havelock-Allan (1830–1897) would later write about Sheridan's tactics at the Battle of Sailors Creek (April 6, 1865; today, Sayler's Creek): "The mode in which Sheridan, from the special arming and training of his cavalry, was able to deal with this rear guard, first to overtake it in retreat, then to pass completely beyond it, to turn to face it, and take up at leisure a position strong enough to enable him to detain it despite its naturally fierce and determined efforts to break through, is highly characteristic of the self-reliant, all-sufficing efficiency to which at this time the Northern horseman had been brought..."

General Ulysses Grant, General-in-Chief of the Union army for the last year of the war, said of Sheridan, "I believe General Sheridan has no superior as a general, either living or dead, and perhaps not an equal."

THE BATTLE AT BRANDY STATION

June 9, 1863—almost a month before the Battle of Gettysburg—Confederate cavalry under General J.E.B. Stuart was at Brandy Station, screening the massive movement of two full infantry corps past Culpepper, Virginia, some six miles southwest of the cavalry location. The infantry destination was the Shenandoah Valley, northward, past Maryland, and into Pennsylvania. General Stuart intended merely to run maneuvers to keep the wary Union reconnaissance guessing.

But the senior officer of the Army of the Potomac, General Hooker, told Union cavalry commander, General Pleasanton, to "break up Stuart's raid in its incipiency." Hooker had mistaken the location and movements of the Confederate cavalry as preparation to disrupt Union supply lines—one of the critical functions of Civil War cavalry. That was not Stuart's purpose; he merely wanted his cavalry to protect the nearby infantry from discovery—another critical function of the cavalry called "screening."

Confederate General Stuart had 9,500 cavalrymen. Little did he know, but 8,000 Union cavalrymen were headed for his location, accompanied by an additional 3,000 infantrymen.

Late afternoon, Stuart's men took a break before continuing north toward Pennsylvania. They were rudely interrupted by General John Buford's First Division cavalry—nearly 5,500 men—rousing the

Confederates so abruptly that some were partly dressed, and many were riding bareback in response to the impending confrontation.

Any discussion of Civil War cavalry would be incomplete without mentioning Union brigade commander Colonel Benjamin Franklin "Grimes" Davis (1831–1863). He was killed in the fierce fighting at Brandy Station. One year earlier, Colonel Davis had escaped under cover of darkness from enemy encirclement at Harper's Ferry. He and 1,300 cavalrymen crossed on a pontoon bridge at night, evading Confederate pickets along the road leading away from the Potomac River. As they worked their way toward safety, Davis and his men stumbled upon a wagon train belonging to Confederate General Longstreet. Using his natural, Mississippi, Southern accent, he ordered the wagon train commander to change directions and to take on his 1,300 men as escorts. In the feeble glow of dawn, the Confederates found the guns of their adopted escorts aimed at them, thus letting them know that they were now prisoners of war.

At Brandy Station, with the death of Colonel Davis, the Union lost a daring and clever cavalry officer. He was a leader and temporary commander of 1st Brigade in General Buford's First Division, Cavalry Corps, while Colonel William Gamble (1818–1866) had been away on medical leave.

A separate, unrelated command had directed Brigadier General David McMurtrie Gregg (1833–1916) and his men to Brandy Station, which led to a second interruption to Confederate General Stuart's plans.

In a confusing sequence of colliding events, General Stuart barely prevailed despite being caught off guard twice in one day. Though his cavalry suffered 575 casualties, the Union had even greater losses at 866.

Brandy Station thus became noteworthy in history as the most significant cavalry conflict in North American history.

CAVALRY AT GETTYSBURG

General Lee had made good use of cavalry in advancing his forces into Pennsylvania and terrorizing the North.

While Pickett's Charge was underway, on the afternoon of July 3,

1863, a cavalry battle was taking place roughly 3 miles to the east. In an area today known as East Cavalry Field, the cavalry division under Jeb Stuart attempted to skirt around the Union's right flank and approach from their rear. Stuart did this to maximize whatever successes could be derived from Pickett's Charge.

Union General David M. Gregg (the second general to interrupt Stuart at Brandy Station) and General George Armstrong Custer (1839–1876) successfully blocked the Confederate assault. Gregg was commander of Cavalry Corps Second Division; Custer was commander of 2nd Brigade under Cavalry Corps' Third Division senior officer Judson Kilpatrick (1836–1881), but detached to assist General Gregg.

After Pickett's Charge was repulsed, in an area southwest of Big Round Top, now known as South Cavalry Field, General Kilpatrick ordered a cavalry charge based on no known reconnaissance or military strategy. Dozens of Union cavalrymen were slaughtered, including his 1st Brigade commander, General Elon J. Farnsworth (1837–1863).

FINAL NOTE

Though these United States had not made the most of cavalry until the Civil War, the next major war would be aided not by horses, but by motorized vehicles—trucks for transporting troops great distances, and tanks to replace the shock charges of saber-wielding cavalrymen. For this brief moment in history, the cavalry, as a tool of war, had become honed to near perfection.

TEST YOUR UNDERSTANDING

Answers to the following questions may be found in the Appendix.

1. Which of the following was not a traditional use of the cavalry?

- *Covering.*
- *Reconnaissance.*
- *Feinting.*
- *Escorting.*

2. What cavalry method did Union General John Buford use west of Gettysburg on the first day?

- *Escorting.*
- *Feinting.*
- *Dismounting.*
- *Shock Charging.*

3. Why was the Union army reticent to adopt full use of cavalry at the start of the war?

- *Most recruits needed to be trained how to ride a horse.*
- *The Civil War was destined to be a short-lived conflict, so the extra effort would be wasted.*
- *Equipping and maintaining a cavalry regiment was expensive.*
- *All of the above.*

4. What was a dragoon?

- *Regular cavalry fighting primarily from horseback and using sabers and pistols.*
- *Support personnel for the cavalry, primarily grooming and feeding horses.*
- *Officers and soldiers using a mix of fighting styles, both dismounted and on horseback, and using carbines, pistols and sabers.*
- *None of the above.*

5. How did General Lee use his cavalry to help augment any successes from Pickett's Charge?

- *Attacking the Union troops from the rear.*
- *Attempting to go around the Union left flank to distract from Pickett's assault.*
- *Utilizing a shock charge to soften the Union line before Pickett's men arrived.*
- *None of the above.*

SUMMING IT UP

During the Civil War, both sides of the conflict developed the cavalry into a superior fighting force, expanding on the functions that had been used here in the past, as well as in Europe.

The cavalry performed several traditional functions such as screening, covering, uncovering, reconnaissance, escorting, raiding, and delivering. They also performed new functions, including General Buford's cavalrymen dismounting to delay the Confederates at Gettysburg. General Stuart attempted to screen Confederate troop movements in the weeks leading up to the battle but was thwarted by the Union cavalry's surprise attack.

Though cavalry played an important role, the culmination of three days of fighting at Gettysburg was the infantry and artillery assault known as Pickett's Charge.

8

PICKETT'S CHARGE

"General Lee... did not give or send me orders for the morning of the third day, nor did he reinforce me by Pickett's brigades for morning attack. As his head-quarters were about four miles from the command, I did not ride over, but sent, to report the work of the second day. In the absence of orders, I had scouting parties out during the night in search of a way by which we might strike the enemy's left, and push it down towards his centre. I found a way that gave some promise of results, and was about to move the command, when he rode over after sunrise and gave his orders."

— LT. GENERAL JAMES LONGSTREET, 1896, *FROM MANASSAS TO APPOMATTOX: MEMOIRS OF THE CIVIL WAR IN AMERICA*, CHAPTER XXVIII, P.385

In the days leading up to Gettysburg, Lee ordered his generals to engage with the enemy only once they had consolidated their forces at Cashtown, Pennsylvania. Some had disobeyed, becoming embroiled in battle with Union General Buford's forces. Rather than turn the situation to Confederate advantage, Lee's generals added disappointment to disobedience by failing to take Cemetery Hill before the enemy could solidify their defenses. On the second day, Lee was disappointed again by the late arrival of his cavalry.

General Lee had invested so much in this effort to force an end to the Civil War, but had been frustrated at nearly every turn. Lee's frustration and disappointment in his generals may have made it difficult to listen to their counsel on that third day of battle.

LEE'S DESPERATE GAMBLE

Modern scholars have differing views about the strategic goal behind General Lee's plan on the third day of the Gettysburg battle. No notes or written orders regarding that goal survived. Some say that Lee wanted to capture the high ground at or near a copse of trees on Cemetery Ridge. Though that group of trees is an impressive cluster today, in 1863, it stood as only a modest grove less than ten feet tall and not visible from many locations on the battlefield of the mid-19th century. Others claim that he ultimately wanted to take Cemetery Hill some 300 yards to the north, never wavering in the plan from the previous day's fighting.

Whatever the ultimate goal, General Longstreet did not like Lee's plan. He thought it foolhardy and futile—an act of mad desperation that would end in far too many deaths and Confederate failure. Years later, Longstreet would write in his memoirs of telling General Lee:

> *"General, I have been a soldier all my life. I have been with soldiers engaged in fights by couples, by squads, companies, regiments, divisions, and armies, and should know, as well as any one, what soldiers can do. It is my opinion that no fifteen thousand men ever arrayed for battle can take that position."*

But Lee would not entertain any other plan; his mind had been set. As soldiers and officers, it was the duty of those under General Lee to follow his commands no matter where those orders led. Of course, Lee himself failed to follow orders instructing him to Vicksburg to fight Union General Ulysses Grant.

Though the "charge" was later named for General George Pickett

(1825–1875), he was only one of three commanders leading the assault. Major General Pickett was one of the division commanders under First Corps' Lieutenant General Longstreet. The other two officers were Brigadier General James Johnston Pettigrew (1828–1863), who led General Heth's division under Third Corps' Lieutenant General Ambrose P. Hill, and Major General Isaac R. Trimble (1802–1888), who led General Pender's division, also under General Hill.

And the "charge" was not entirely a charge, either. Most of the assault would be a methodical, steady march across largely open fields on an undulating terrain, with an actual charge during only the last few dozen yards.

General Pettigrew led four brigades commanded by:

- Colonel Birkett D. Fry (1822–1891; Archer's Brigade),
- Colonel James K. Marshall (1839–1863; Pettigrew's Brigade),
- Brigadier General Joseph R. Davis (1825–1896), and
- Colonel John M. Brockenbrough (1830–1892; Heth's "old" Brigade).

General Trimble led two brigades commanded by:

- Colonel William Lee J. Lowrance (1836–1916; filling in for wounded General Scales) and
- Brigadier General James Henry Lane (1833–1907), not to be confused with Union General James Henry Lane (1814–1866), who had also been a sitting United States senator [R-KS] while wearing his uniform.

Because General Hill had fallen ill, Pettigrew's and Trimble's divisions were under General Longstreet's command during the assault.

An additional two brigades from General Hill's Third Corps, Anderson's division, stood ready to assist from the Confederate right flank, led by Brigadier General Cadmus M. Wilcox (1824–1890) and Colonel David Lang (1838–1917; Perry's brigade).

Initially, the Confederate assault was aimed squarely at Union General Winfield Hancock's II Corps. And in the center of Hancock's portion of the Union line stood Brigadier General John Gibbon (1827–1896) and his Second Division of II Corps. General Meade warned Gibbon during the previous night's planning that this was likely to happen. Under General Gibbon were:

- 1st Brigade under Brigadier General William Harrow (1822–1872),
- 2nd Brigade under Brigadier General Alexander Webb (1835–1911), and
- 3rd Brigade under Colonel Norman J. Hall (1837–1867).

While Pickett, Trimble, and Pettigrew were busy approaching the Union front, J.E.B. Stuart's cavalry was to occupy the Union forces from the rear. Stuart was to disrupt communication lines between the various parts of the Union line, and to disrupt any anticipated retreat, which would most likely have occurred on the southbound Baltimore Pike.

Any general in their right mind would have wanted an early start to make the most of the daylight hours, especially the cooler hours of the morning since it would be Confederate troops on the move. Later, critics of the Confederate performance at Gettysburg would blame Longstreet for not ensuring the men were ready earlier in the day.

Despite the fumbled opportunity to the west of the Union position, a long, hard battle was being fought at Culp's Hill at the far-right end of the Union line. If Longstreet had been more proactive, he could have helped to apply far more pressure on General Meade by simultaneously attacking a second major front. By delaying Pickett's Charge, the overall effect fizzled.

The Union army had two key advantages: superior numbers, even after all of the losses of the first two days; and the coveted high ground.

If Pickett's Charge had begun at the same moment as the assault on Culp's Hill, General Meade would have needed to divide his forces and monitor carefully the possible need to shift one or more regiments to help where one part of the line was being lost.

However, in the real world of July 3, 1863, Longstreet's cannon assault began just about the time that Confederate forces under Major General Edward "Allegheny" Johnson were losing momentum.

CANNON BARRAGE

Starting about 1 p.m. and for the next two hours—in perhaps the most significant artillery engagement of the entire war—cannon from both sides filled the air with smoke and shells.

The objective was simple for the Confederates: concentrate enough firepower at one location on the Union line to destroy it and demoralize the remaining Union infantry.

Colonel Edward Porter Alexander (1835–1910) was in charge of Longstreet's artillery brigade on the third day at Gettysburg. General Lee's Chief of Artillery, General William Pendleton (1809–1883), however, apparently did more to hamper Alexander's efforts than to help. Artillery brigades from other divisions did not coordinate well with Alexander. The overall effect was a diffuse assault in the general direction of the Union line rather than a concentrated artillery barrage, frequently overshooting the line. Even so, a sufficient number of shells pelted General Meade's headquarters at the Leister farmhouse that he had to have his staff move to safety. One photograph after the battle showed several of Mrs. Leister's horses dead in the yard from the artillery barrage.

The Confederate forces used about 160 guns distributed throughout a line more than two miles long from the vicinity of the Peach Orchard, northward, roughly parallel to Emmitsburg Road.

Confederate Brig. Gen. Evander M. Law later wrote:

"The cannonade in the center ... presented one of the most magnificent battle-scenes witnessed during the war. Looking up the valley towards Gettysburg, the hills on either side were capped with crowns of flame and smoke, as 300 guns, about equally divided between the two ridges, vomited their iron hail upon each other."

Despite the intensity of the artillery bombardments, the overall effect was relatively weak. Some shells detonated late or not all due to substandard fuses. In addition, the amount of smoke clouding the view made it difficult for the gunners to assess their accuracy.

On the Union side, Brigadier General Henry J. Hunt (1819–1889), General Meade's artillery chief, had roughly 80 cannon at his disposal to counter the Confederate onslaught. Some placement locations were not optimum, but good options were scarce. He was also quickly running out of ammunition in the heated exchange. To conserve his ammunition for more strategic use and to trick his enemy into thinking they had won the artillery barrage by destroying most of his guns, Hunt had his men reduce the cannon fire at a sufficiently slow rate that he may well have accomplished this deception.

But the commander of II Corps, Major General Hancock, insisted on more artillery fire to keep the morale high for the infantrymen being assaulted by Confederate cannon. Thus, Hunt's plan was overridden, and he ran out of ammunition before Pickett's Charge. This, he would later reflect, led to an inability to thrash the Confederate advance, which could have forced them to reconsider. Hunt's plan could have saved numerous Union lives.

LONGSTREET'S ASSAULT—THE CHARGE THAT WASN'T

By the time the infantry assault began, the day had become uncomfortably hot and humid—an estimated 87°F (31°C). Though the Union cannon targeted the opposing artillery, they frequently aimed too high, often slamming into the awaiting infantry.

General Longstreet despised Lee's plan; perhaps that is why he did not promptly order the advance. Instead, he told Colonel Alexander to choose what he thought might be the right time to start and then to inform General Pickett. Alexander wrote to Pickett:

> *"If you are coming at all, come at once, or I cannot give you proper support, but the enemy's fire has not slackened at all. At least eighteen guns are still firing from the cemetery itself."*

Only a short time afterward, Pickett asked Longstreet if he should begin. His commander later wrote in his memoirs:

> *"The effort to speak the order failed, and I could only indicate it by an affirmative bow."*

The uncertainty Longstreet felt had become palpable. Once Pickett left with Longstreet's nod to begin, the commander sought out Colonel Alexander to assess the current status of the artillery. With their ammunition nearly exhausted, Alexander explained that his guns might not be able to offer much support during the assault. Yet, when Longstreet told Alexander to reverse his order to Pickett, Colonel Alexander informed the general that resupplying with ammunition from the trains would take too long, destroying any advantage gained from the lengthy barrage they had already done.

Pickett, Trimble, and Pettigrew ordered their men forward without the planned support of a continuing artillery barrage against their enemy.

It was about 2 p.m. when about 12,500 Confederate infantrymen began the methodical march toward the Union line. From the Confederate viewpoint, the combined forces of Pettigrew and Trimble stood on the left, while Pickett's men were arrayed to the right—a total of nine brigades made up a solid line of men over a mile long.

Right from the start, the Confederates faced renewed artillery bombardments and several fences blocking their way. In all, they had to cross nearly 1,300 yards of dangerous no-man's land.

Union artillery fire came from three different directions:

- A concealed location north of Little Round Top, directed by Lieutenant Colonel Freeman McGilvery (1823–1864) against the right flank of the Confederate forces.
- Cannon placed atop Cemetery Hill pounded the Confederate left flank.
- II Corps artillery reserve on Cemetery Ridge.

After negotiating the lethal terrain, the last, relatively safe haven on the long march of death stood about 300 yards from the Union line at Emmitsburg Road. Confederate morale had begun to plummet, and unit cohesion had all but disappeared. Several thousand infantrymen became glued to the virtual safety at the edge of the road and refused to move any closer to certain death. Many of them were later taken as prisoners of war once the battle ended.

Of those Confederate soldiers who continued, their offensive line had been reduced to less than 900 yards in length, skittishly attempting to avoid the flanking artillery fire.

Confederate Colonel John Brockenbrough (1830–1892) led the northernmost brigade against the Union army. He commanded General Heth's old brigade before Heth had been made division commander. At the edge of the Confederate left flank, they received the heaviest assault from the artillery atop Cemetery Hill. When they came within musket range, they were greeted by a powerful rain of bullets as the 8th Ohio regiment established a barrage of fire so intense that the Confederate troops panicked and retreated to Seminary Ridge. As they clawed their way through Trimble's division, the terror of Brockenbrough's men infected others also to turn and run.

With the success of their barrage, the Ohio regiment launched their flanking assault against the new Confederate left flank led by Brigadier General Davis.

Those of Pettigrew's division who made it past Emmitsburg Road were hit by another volley masterfully directed by Union General Alexander Hays (1819–1864). Secure behind 260 yards of stone, riflemen from his division were arrayed in lines, sometimes four deep, each line stepping up to the stone wall, firing, and then moving back to reload for a near-constant barrage of musket fire.

Although the Union line was also receiving heavy fire from the approaching Confederates, General Hays rode back and forth, cheering on his men despite the personal danger. And despite losing his first two horses, the general continued to show bravery to his men.

Union commander of II Corps, General Hancock, also displayed bravery to his men by riding back and forth despite making himself an easy target. About this time during Pickett's Charge, Hancock was struck in the thigh by a bullet that bounced off his saddle's pommel. Despite the injury, General Hancock insisted on staying with his men until the end of the current engagement.

On General Pickett's side of the assault, the Confederate right flank began rotating northward until the line faced toward the northeast. Two lines marched in this new direction, including the brigades led by

- Brigadier General James L. Kemper (1823–1895), far right flank,
- Brigadier General Richard B. Garnett (1817–1863), on the left of Pickett's line, and
- Brigadier General Lewis A. Armistead (1817–1863) followed close behind.

This maneuver unavoidably exposed Kemper's men to concentrated fire because this gave the Union guns of Colonel McGilvery and rifles of General Doubleday's division on Cemetery Ridge a narrower field of focus. To make matters worse for Kemper's brigade, Union I Corps, Third Division, 3rd Brigade, led by Brigadier General George Stannard (1820–1886), realizing their opportunity, moved out onto the field of battle to come up behind the exposed, Confederate brigade, and repeatedly fired into the backs of the enemy.

As Pickett's division angled toward the center of the Union line, they were assaulted by a rain of heavy fire from one brigade after another.

When Confederate forces finally reached the Union line, at a salient called the "Angle," one commander there ordered his regiment (71st Pennsylvania) to retreat, leaving the 1st New York, an artillery battery, to fend for themselves. This left a dangerous gap in the Union line for the nearly 3,000 approaching enemy men.

Fierce fighting with bayonets, rifle fire, and desperate hand-to-hand combat ensued. As the Confederates seemed to be gaining the upper hand, they seized two artillery guns and turned them on the nearby Union soldiers. But their sacrifices and accomplishments fell apart when they realized that there was no longer any ammunition for the guns. Moments later, the 72nd Pennsylvania arrived to plug the hole made by the 71st.

Nearly all of the Confederate commanders had been killed, so as the Union infantrymen swarmed toward their position, none of the remaining Southerners had sufficient authority to organize a safe retreat.

THE COST OF PICKETT'S CHARGE

Pickett's Charge had taken only minutes shy of an hour. The Union Army of the Potomac suffered some 1,500 men who were either killed or wounded. Confederate losses were much higher. For Pickett's division:

- 498 were killed,
- 643 were wounded (and not captured),
- 833 were wounded and captured, and
- 681 were captured (and not wounded).

Pettigrew's losses amounted to similar numbers:

- 470 killed,
- 1,893 wounded, and
- 337 captured.

Trimble's losses were somewhat less:

- 155 killed,
- 650 wounded, and
- 80 captured.

Additional losses included:

- 200 from Wilcox's brigade, and
- 400 from Lang's brigade.

Over half of Pickett's officers of field grade were killed, wounded, or captured. As the Confederate infantry that remained returned to the relative safety of Seminary Ridge, General Lee ordered General Pickett to prepare his division for a Union counteroffensive. Pickett replied to his commander: "General, I have no division."

After an exhausting three days, General Meade was grateful merely to have kept the high ground and to have stopped the Confederate assaults.

The following day—the North's sacred Independence Day—an

unofficial truce allowed men from both sides to gather their dead and wounded.

In all, the Union victory at Gettysburg, plus General Grant's victory over Vicksburg, Mississippi, significantly shifted the course of the war between North and South. After two full years of war, the advantages of the Confederacy—the skills with guns and horses—were being replaced with the advantages of the Union—superior numbers, greater manufacturing capabilities, and a more considerable depth of resources.

General Pickett's after-battle report contained so much animosity that General Lee had him remove it from the record. Some historians have found it ironic that General Pickett should survive when so many of his brigade and regiment officers did not. The standard rule of warfare throughout most of history has been one of the senior commanders leading from the rear, with junior officers leading from the front. To be sure, there have been notable exceptions, but General Pickett was following the convention of the times.

TEST YOUR UNDERSTANDING

Answers to the following questions may be found in the Appendix.

1. Which three Confederate commanders led the charge across the battlefield?

- *General George Pickett, General Ambrose Hill, General Isaac Trimble.*
- *General George Pickett, General James Longstreet, General James Pettigrew.*
- *General George Pickett, General Isaac Trimble, General James Pettigrew.*
- *None of the above.*

2. Why were all three divisions for the charge placed under General Longstreet's command?

- *President Jefferson Davis had requested an update, and General Lee sent General Ambrose Hill in his place.*
- *General Ewell was ill and couldn't take command.*

- *Generals Pettigrew and Trimble needed to get along with their commander.*
- *None of the above.*

3. What made the Confederate efforts on the third day fizzle?

- *General Longstreet not preparing his troops for an assault first thing in the morning.*
- *General Ewell starting too early with his attack at Culp's Hill.*
- *General Stuart making too much noise as he attempted to circle the Union right flank.*
- *None of the above.*

4. How did the assault managed by General Longstreet begin?

- *By attacking Culp's Hill early in the morning.*
- *With a loud "Whoop!" by the Confederate infantrymen to terrify the Union line.*
- *By sending General Pickett's infantry to do a flanking maneuver at the Union left.*
- *With an artillery barrage lasting one to two hours starting early afternoon.*

5. How many Confederate infantrymen stepped off toward the Union line as part of Pickett's Charge?

- *34,800.*
- *12,500.*
- *17,900.*
- *8,500.*

SUMMING IT UP

Pickett's Charge was neither Pickett's nor a full charge. In hindsight, General Longstreet's disagreement with Lee's plan appears to have been justified.

The barrage and counter-barrage of artillery fire did not soften up either side. And though the Union artillery chief had wanted to hold some ammunition in reserve to stop the charge in its tracks, his desires were overridden by the need to keep up morale.

The third day of the Battle of Gettysburg could have turned out differently if the coordination between corps and divisions had been more streamlined.

So many men died crossing the first few hundred yards of the battle-field that morale all but evaporated. When General Pickett returned to Seminary Ridge, the Battle of Gettysburg was over. All that remained was to assess the damages and plan the next desperate move.

9
GETTYSBURG AFTERMATH

"So far from engaging in a war to perpetuate slavery, I have rejoiced that slavery is abolished. I believe it will be great for the interests of the south. So fully am I satisfied with this, as regards Virginia especially, that I would cheerfully have lost all I have lost by the war, and have suffered all I have suffered, to have this object attained."

— ROBERT E. LEE, STATEMENT TO JOHN LEYBURN
(MAY 1, 1870), AS QUOTED IN *R. E. LEE: A
BIOGRAPHY*, 1934, BY DOUGLAS SOUTHALL
FREEMAN

THE TOLL OF GETTYSBURG

The Battle of Gettysburg contained a series of mistakes, disobedience, and miscalculations. More than that, however, Gettysburg was one of the bloodiest battles in human history, especially in lives lost per day.

If President Jefferson Davis had not launched an unprovoked attack on Fort Sumter, the Civil War may never have happened. If the Confederate generals had followed Lee's orders not to engage with the enemy until they had assembled at Cashtown, Pennsylvania, perhaps the

South would have won their inevitable battle in Pennsylvania. If only Lee's Cavalry Corps had joined him on July 1, he could have had the advantage of his cavalry's reconnaissance skills. If General Ewell had only pushed to take Cemetery Hill on the evening of July 1, Union General Meade may have been forced to withdraw from Gettysburg. Suppose only Union artillery chief General Hunt had been allowed to use his ammunition more strategically. In that case, he might have been able to save Union lives by blasting the Confederate advance right at the start of their dangerous progress from Seminary Ridge.

Of those killed on the Union side, the following counts are for their number by corps:

- I Corps—666
- II Corps—797
- III Corps—593
- V Corps—365
- VI Corps—27
- XI Corps—369
- XII Corps—204
- Cavalry Corps—91
- Artillery Reserve—43
- Total—3,155

Those who were wounded on the Union side:

- I Corps—3,231
- II Corps—3,194
- III Corps—3,029
- V Corps—1,611
- VI Corps—185
- XI Corps—1,922
- XII Corps—812
- Cavalry Corps—354
- Artillery Reserve—187
- Total—14,525

The number killed on the Confederate side:

- First Corps—1,617
- Second Corps—1,301
- Third Corps—1,724
- Cavalry Corps—66
- Total—4,708

And the number wounded on the Confederate side:

- First Corps—4,205
- Second Corps—3,629
- Third Corps—4,683
- Cavalry Corps—174
- Total—12,691

These are the best counts available, but overall totals vary from source to source. Including soldiers who ended up missing or captured, the casualties for the Union side are estimated at 23,000, while those for the Confederacy have been estimated at 28,000—a total of 51,000. One author has proposed as many as 57,225 casualties for the total.

LEE'S RETURN TO VIRGINIA

On July 4, 1863, heavy rain washed away some of the blood from the now-silent fields between the two armies. Lee had arrayed his forces on Seminary Ridge into a defensive posture, hoping Meade would attack. However, the Union general did not take the bait and was later criticized heavily.

Toward the middle of the continent, General Ulysses Grant had forced a surrender from the Confederate garrison at Vicksburg, Mississippi, effectively splitting the South into two regions.

Instead of fighting, both armies began to collect their dead and wounded. Hoping for some compassion, Lee sent a request to Union General Meade, asking if he would agree to a prisoner exchange. Meade rejected the offer.

Amidst the dreary rain, Lee began moving his support personnel, wounded, and materials back to Virginia, keeping only the fighting force close at hand. The non-fighting part of his army consisted of a wagon train that stretched for seventeen miles and would work its way past Cashtown to Williamsport, Maryland, and on to the Confederacy.

When evening came, Lee's fighting force took a more direct route past Fairfield but through far more rugged terrain.

Late on July 4, General Kilpatrick made up for his blunder the day before by capturing roughly 200 or more wagons and close to 1,400 prisoners.

General Meade was faced with a different set of challenges. The simplest was that the army that left the battlefield first was considered the loser, but this was obvious by the morning of July 5. However, more importantly, Meade needed to ensure that Lee did not gain direct access to either Baltimore or Washington.

Two days later, another battle took place in Maryland, not far from the Potomac River. The weather had stopped General Lee and his men. The rain caused the water to rise in the flowing river, destroying the pontoon bridge Lee intended to use.

General Meade moved his infantry to pursue the Confederate army three days after Lee left Gettysburg. General Henry Halleck (1815–1872), the General-in-Chief of the entire Union army, tried numerous times to get Meade to be more aggressive in chasing after Lee's army. President Lincoln similarly implored Meade on this.

By July 13, construction of a new pontoon crossing had been completed. Although Meade's infantry had arrived in the area on the previous day, it was Meade's cavalry that moved on the Confederate army's rear guard on the day following their crossing. Meade took some 500 prisoners. During the skirmish, General Pettigrew was wounded and died three days later.

After another nine days, on July 23, the two armies clashed again at the Battle of Manassas Gap. General Lee escaped capture, and General Meade decided to stop the pursuit.

REACTIONS TO THE NEWS

Union

After having struggled through the first two years of the Civil War, and with the South making the best of their few advantages, the news of Lee's defeat at Gettysburg led to claims that the crushing blow to the

South had "eclipsed" Napoleon's loss at Waterloo. One New York writer said in his *Diary,* "The charm of Robert E. Lee's invincibility is broken," and "Government is strengthened four-fold at home and abroad" (George Templeton Strong).

Public optimism did not last long when they realized that Meade had let Lee's army slip through his fingers, thus prolonging the war. This is the very problem Lincoln had hoped to solve by selecting Meade to replace General Hooker.

President Lincoln expressed frustration in writing to Navy Secretary Gideon Welles, "Our army held the war in the hollow of their hand and they would not close it!"

The president was not alone in his dissatisfaction. Commander of II Corps, Second Division, 2nd Brigade, Brigadier General Alexander Webb, wrote in dismay to his father on July 17—two weeks after Gettysburg—that several politicians in Washington, D.C. had contacted him. These included "Chase, Seward and others," telling him how annoyed they were with Meade—enough to "write to [Webb] that Lee really won that Battle!"

There were 72 Medals of Honor issued to those who fought either directly in the Battle of Gettysburg (64) or in the actions leading up to or directly following the primary battle (8). Seventeen were issued during the Civil War, mainly in December 1864. Most Medals of Honor were issued during the 1890s when most Civil War survivors had entered late middle age or their senior years.

Confederacy

While the battle was still underway on July 3, Confederate President Jefferson Davis sent his Vice President, Alexander Stephens, with a truce flag toward the Union forces at Norfolk, Virginia, some 147 miles (237 kilometers) south of Washington, D.C. His mission had been one of negotiating an exchange of prisoners. Davis fully expected that General Lee would soon threaten the Union capital city from the north, while Stephens approached it from the south.

By the time Lincoln received Stephens's request to approach Washington for negotiations, the Union president had received news of the final Gettysburg results. Lincoln had no incentive to negotiate and refused to allow the Confederate vice president's request.

Weeks later, when news of Gettysburg reached London, the Confederate hopes of European recognition were destroyed. The South would not be receiving European aid for their cause.

The combination of Lee's loss at Gettysburg with Grant's win at Vicksburg made it clear to those watching from afar that the Confederacy had little hope of winning. In addition to the massive losses of men at Gettysburg, the South lost another 30,000 men, plus ammunition and weapons, with the fall of their Mississippi fortress at Vicksburg.

However, those in the South tried to paint Gettysburg as merely a temporary setback. They attempted to use every positive aspect of General Lee's Pennsylvania Campaign to create the impression that the battle had been successful. For one, Virginian farms were spared a summer of abuse.

Even General Lee tried to view the loss at Gettysburg positively. He wrote to his wife that he had come back to Virginia "rather sooner than I had originally contemplated, but having accomplished what I proposed on leaving the Rappahannock, viz., relieving the Valley of the presence of the enemy and drawing his Army north of the Potomac." Lee reportedly said to Major John Seddon, whose brother was the War Secretary of the Confederacy, "Sir, we did whip them at Gettysburg, and it will be seen for the next six months that that army will be as quiet as a sucking dove." Yet, General Lee was not as cocky when facing his president on August 8, 1863—more than five weeks after Gettysburg. Instead of bragging, Lee attempted to resign his commission. President Davis rejected the offer.

TEST YOUR UNDERSTANDING

Answers to the following questions may be found in the Appendix.

1. How many were estimated to have been killed at Gettysburg, including both Union and Confederate armies?

- *7,863.*
- *14,525.*

- *12,691.*
- *51,000.*

2. What did General Lee hope on the fourth day at Gettysburg?

- *That General Grant would realize that he had lost.*
- *That General Meade would go home defeated.*
- *That President Lincoln would recognize the futility of dislodging Lee from his strong position and sue for peace.*
- *That General Meade would attack his strong, defensive position on Seminary Ridge.*

3. What was accomplished on July 4, 1863, at Gettysburg?

- *Lee's noncombatant forces returned to Virginia unmolested.*
- *Both sides gathered their wounded and some of their dead.*
- *Union forces set off fireworks to celebrate Independence Day.*
- *All of the above.*

4. On July 3, General Kilpatrick had unthinkingly attacked the Confederate right flank, losing much of his cavalry with no clear strategy; how did he make up for this blunder on July 4?

- *Personally took care of burying all the men he lost.*
- *Captured 200 wagons and 1,400 prisoners.*
- *Forced General Lee to flee from his stronghold on Seminary Ridge.*
- *All of the above.*

5.W hy did General Meade not pursue Lee's fleeing army right away?

- *He wanted to ensure that the Confederate army could not attack Washington, D.C.*
- *The first to leave a field of battle is traditionally considered to be the loser.*
- *Neither of these.*
- *Both of these.*

SUMMING IT UP

The Battle of Gettysburg was punctuated by errors borne of human fallibility and hubris. The chief legacy was one of immense bloodshed and too few lessons learned.

Another result of Gettysburg involved the eventual outcome of the Civil War, for though no war turns on a single event, the Battle of Gettysburg played an oversized role in that outcome. Even so, General Meade could have shortened the war by months, if not an entire year, simply by attacking his foe more aggressively—a failing for which his predecessor had been removed as commander of the Army of the Potomac.

Both the Union and the Confederacy tended to mischaracterize what happened. But both generals betrayed their side by losing at least in one fundamental way. Lee disobeyed orders to help at Vicksburg, gambling on his idea of ending the war early. Meade let the opportunity slip away to end the Civil War within days, if not hours.

The Civil War, like nearly every other battle throughout history, would find the most significant suffering in the care afforded to the wounded.

10

MEDICAL CHALLENGES

"After seven o'clock a silence began to descend on the whole field. Then, along the crest, the men could hear the cries and moans of the wounded below them. The Confederate medical corps were always understaffed, and, because medicines were shut off by the blockade, the doctors and medical volunteers always worked without proper drugs. In amputations without anesthesia, screams would be torn from even the most courageous men, and these agonized sounds scraped across the nerves of the soldiers more harshly than the whine of the largest shells."

— CLIFFORD DOWDEY, 1958, *LEE AND HIS MEN AT GETTYSBURG: THE DEATH OF A NATION,* CHAPTER 8

THE STATE OF MEDICINE DURING THE CIVIL WAR

Many needless deaths can be attributed to the poor state of medical knowledge during the 1860s. Procedures we take for granted today were unknown to the medical professionals of the mid-19th century, even in Europe. Simple things, like hygiene, sanitation, antiseptic surgery, and sterile dressings, were not considered by Civil War doctors.

Medical schools of the time graduated their students after two years of study. Most doctors on the battlefield had never before dealt with such wounds. Because doctors were wholly ignorant of the reasons why so many died from their wounds, they frequently opted to amputate rather than risk the seemingly inevitable infections from which survival rates were meager.

The nursing care developed by British medical worker Florence Nightingale during the Crimean War (1853–1856) remained unknown to most American doctors. All too often, a battlefield wound was a death sentence. Those who survived their wounds suffered great pain without the modern medicines available today.

A far deadlier problem for Civil War soldiers involved disease. The unsanitary conditions frequently led to ailments like diarrhea and dysentery. Others succumbed to malaria, pneumonia, smallpox, syphilis, typhoid fever, and tuberculosis. And though not necessarily deadly, childhood illnesses like measles and mumps left many soldiers debilitated.

Country boys who had never contracted such simple illnesses and had no immunity were crowded into tents. Poor sanitation, combined with inadequate food and water, led to significant loss of life—as much as twice the losses from battlefield injury.

Because of the emergencies of war, sometimes recruiting require-ments were relaxed to gain more bodies in uniform. Simple examina-tions were not done to weed out those who were diseased.

Antiseptic techniques developed by British doctor and surgeon Joseph Lister (1827–1912) were not discovered until August 1865, months after the end of the Civil War. Moreover, it took years for Dr. Lister to perfect his techniques. The first several years after his initial discoveries were filled with ridicule from his peers. The honors he received came later in life.

The doctors of the Civil War used mainly amputation. However, as the war progressed, doctors were forced to become more organized. One of their earlier developments included better record-keeping to under-stand what worked and what did not. Afterward, the dissemination of the beneficial procedures helped more doctors improve their care of the ill and wounded.

Union Army surgeon general and Brigadier General William A. Hammond developed a different kind of army hospital that was cleaner and better ventilated, resulting in a death rate that plummeted to only 8%. The use of stretchers and ambulances, plus morphine for extreme cases of pain, allowed more of those wounded on the battlefield to survive.

TEST YOUR UNDERSTANDING

Answers to the following questions may be found in the Appendix.

1. What medical procedure missing during the Civil War could have improved survival rates?

- *Using nurses to improve the morale of the wounded soldiers.*
- *Disinfecting surgical tools before performing surgery.*
- *Entertainment to distract patients from their pain.*
- *None of the above.*

2. What effects did poor sanitation cause?

- *Increased rates of infection to existing wounds.*
- *Spread of diseases amongst patients.*
- *Dysentery and diarrhea from dirty hands and dirty utensils.*
- *All of the above.*

3. When were Dr. Lister's antiseptic techniques discovered?

- *A few months after the Civil War ended.*
- *A decade later.*
- *Twenty years later.*
- *At the turn of the next century.*

4. What practice helped in improving patient care?

- *Entertainment for the patients.*

- *Segregating patients by state.*
- *Keeping better records.*
- *Refrigeration to keep the patients cool.*

5. Using the technique mentioned in question #4, what was the death rate?

- *2%*
- *8%*
- *23%*
- *31%*

SUMMING IT UP

Medical science in the 1860s was only then coming out of the "Dark Ages" of uncertainty and blind tradition. The surgeon general of the Union army, Brigadier General William Hammond, helped to advance medicine significantly by using better records to study what worked and then by implementing those successful actions more broadly.

Of course, the soldiers suffered, especially the wounded, but so did the civilians.

11

GETTYSBURG CIVILIANS

"In the weeks following the battle, Lydia replaced the siding on the barn and house and had the well re-dug. In order to acquire some of the money necessary to finance these repairs, she began selling the bones of dead horses on her property after the meat rotted off them one year later. Although the remains of horses were used for a variety of products, the main usage for bones was to harvest a substance called collagen and use it to make an adhesive. This process must have been extremely appalling and gruesome work for Lydia and her daughters; however their willingness to defy traditional 19th-century gender norms provides an illustrative example of how wartime necessity could, in many instances, stretch and shift the boundaries between masculine and feminine spheres."

— JESSICA ROSHON, APRIL 5, 2021, "SURVIVING, PERSEVERING, AND PROFITEERING: THE STORY OF LYDIA LEISTER AT GETTYSBURG," *THE GETTYSBURG COMPILER*

LIFE IN A WAR-TORN TOWN

At the time of the battle, the population of Gettysburg was about 2,400; today, it is a little over 7,100. The town had been settled in 1780, near the end of the Revolutionary War, and incorporated in 1806. Though young by European standards, the community had become well-established by the time the Civil War had started.

The town is near the southern border of Pennsylvania, only about seven and a half miles (twelve kilometers) from Maryland. For context, Big Round Top was about three miles (five kilometers) south of the center of town.

Houses all along the southern edge of town were riddled with bullets, and the barrages of artillery and musket fire destroyed some structures just outside of town.

There were two civilian casualties in all of the fighting—a twenty-year-old who had been preparing to bake bread for the Union troops nearby and a sixty-nine-year-old volunteer who had previously served in the War of 1812.

After suffering the nerve-wracking noise of battle for three days, the people of the small town were burdened with the stench of close to 8,000 dead soldiers and more than 3,000 dead horses rotting in the hot summer sun. The putrid smell caused some in the town to become fiercely ill.

Besides the burdensome task of burying the dead soldiers and burning the horse carcasses, the two thousand civilians had to assist with the care of wounded Union troops numbering 14,000 and Confederate prisoners of another 8,000 men.

LYDIA LEISTER'S FARM AND GENERAL MEADE'S HEADQUARTERS

Lydia Leister, a 52-year-old widow, had been warned on July 1 that there would likely be fighting on her land, so she took the two daughters who still lived with her to a safer location. The following day, because of its central location within the arcing Union line, General Meade selected the house as his headquarters. On the evening of July 2, her tiny house was the meeting room for a dozen Union generals to plan the fighting the following day.

On July 3, during the bombardment before Pickett's Charge, Mrs. Leister's house fell victim to numerous artillery overshoots and countless bullet holes, forcing General Meade and his staff to leave.

When Mrs. Leister returned home, she found an incredible array of damage:

- Her house was in dire need of repair, including artillery shell damage to her porch supports.
- Seventeen dead horses littered her yard.
- Her prize peach tree had been killed by someone burning dead horses next to it.
- Both her horse and cow were missing.
- All of her apple trees had been destroyed.
- Two tons of hay she had stored in her barn were gone.
- Her wheat field had been trampled beyond use.
- Fencing rails—all of them burned.
- Her water spring had been ruined with the bodies of dead horses.

Mrs. Leister never received any compensation for using her home or the damage her farm had suffered. She did tell a reporter, however, in 1865, that she had been able to make a little money by selling the skeletons of the dead horses.

WOUNDED CIVILIAN VOLUNTEER

John L. Burns (1793–1872) had served in the War of 1812 and the Mexican-American War. When the Civil War started, his attempt to enlist was rejected because of his age.

When the war came to his hometown, he grabbed his old flintlock musket, along with his powder horn, and made his way to the fighting taking place on the morning of July 1, west of town. Upon finding a wounded Union soldier, he requested the use of the soldier's more up-to-date rifle and ammunition. With these in hand, he made his way forward until he encountered Major Thomas Chamberlin (1838–1917), an officer of the 150th Pennsylvania. The old man's request was simple —to fall in with the other soldiers of the regiment—a part of the 2nd Brigade, Third Division, I Corps.

Chamberlin sent Burns to the regimental commander, Colonel Langhorne Wister (1834–1891), who, in turn, sent the senior citizen to McPherson Woods (also called Herbst Woods), thinking that the old man would find suitable shelter out of the way of army business.

Instead of laying low, Burns fought first with the 7th Wisconsin and later with the 24th Michigan—a part of the Iron Brigade. He was wounded several times, and when the Union was forced to fall back, they had to leave Burns behind.

Tired from hours of fighting and weak from his wounds, Burns crawled to put some distance between himself and his borrowed rifle. When the Confederates arrived, he lied, telling them that he had been wandering the battlefield to find help for his ailing wife.

Typically, anyone fighting as an enemy out of uniform was to be executed on the spot. Because of his quick thinking, he convinced the Rebels that he was a noncombatant. Accordingly, their surgeon treated his wounds. That evening, he crawled into the cellar of the closest house. Later, he acquired help to find his way back home and to gain further treatment from the local doctor.

CIVILIAN FATALITY

There was only one known civilian death from the Gettysburg battle. On July 1, 1863, Jennie Wade, age 20, her mother, and two brothers left the relative security of their home in the center of Gettysburg. She traveled to the home of her sister, Georgia Anna Wade McClellan. There, they would assist Georgia and her newborn child. The McClellan house was at the southern edge of town and near the northern end of the Union line, where Baltimore Pike and Emmitsburg Road intersect.

At around 8 a.m. on July 3, a stray bullet went through the kitchen door and pierced young Jennie through the heart, killing her instantly. Jennie's mother heard her drop to the floor and, upon discovering what had happened, immediately told her sister Georgia that Jennie was dead. Two Union soldiers who had been upstairs came down to investigate the women's screams.

Using a coffin initially intended for Confederate General William Barksdale, Jennie's body was buried behind the house and months later moved to her final resting place.

During the three-day battle, the McClellan house had received close to 150 bullet holes.

On July 4, the day after Jennie's death, her mother baked bread with the dough Jennie had been kneading. In all, she made fifteen loaves, most of it for the Union soldiers.

Nineteen years later, the United States Senate voted to give Mrs. Wade, Jennie's mother, a pension for her loss because her daughter had been preparing bread for the Union cause.

TEST YOUR UNDERSTANDING

Answers to the following questions may be found in the Appendix.

1. What was the approximate population of Gettysburg during the battle?

- *370.*
- *2,400.*
- *19,300.*
- *76,100.*

2. Approximately how many rotting corpses did the townspeople have to handle in the wake of the battle?

- *580.*
- *1,500.*
- *8,000.*
- *51,000.*

3. Approximately how many wounded soldiers did the townspeople have to help with?

- *950.*
- *2,800.*
- *14,000.*

- *57,500.*

4. How did Lydia Leister recoup some of her losses after General Meade had used her home as his Gettysburg headquarters?

- *She sold the bones of dead horses.*
- *She charged money for tours of her home.*
- *She created a bed and breakfast for Gettysburg tourists.*
- *All of the above.*

5. How was Jennie Wade made the only civilian fatality of the battle?

- *She was trampled by horses fleeing Pickett's Charge.*
- *She was crushed by the ceiling falling in on her when an artillery shell hit the house.*
- *She was hit by a stray bullet at the start of hostilities on the third day.*
- *None of the above.*

SUMMING IT UP

Though virtually every one of the citizens of Gettysburg survived the battle, they suffered from the noise and fright of war being waged so close to them. They also suffered the sickening smell of thousands of rotting corpses in the hot summer sun. And they suffered the burden of helping to manage thousands of wounded and thousands of prisoners —almost ten times the population of the town.

Gettysburg's quiet, rural lifestyle would once again be upset when a world-renowned orator and the United States president visited the town to commemorate their Civil War cemetery.

12

LINCOLN'S ADDRESS AT GETTYSBURG

"We shall then see that, tinder God's providence, every sheet of flame from the blazing rifles of the contending armies, every whizzing shell that tore through the forests at Shiloh and Chancellorsville, every cannon-shot that shook Chickamanga's hills or thundered around the heights of Gettysburg, and all the blood and the tears that were shed are yet to become contributions for the upbuilding of American manhood and for the future defence of American freedom. The Christian Church received its baptism of Pentecostal power as it emerged from the shadows of Calvary, and went forth to its world-wide work with greater unity and a diviner purpose. So the Republic, rising from its baptism of blood with a national life more robust, a national union more complete, and a national influence ever widening, shall go forever forward in its benign mission to humanity."

— GENERAL JOHN B. GORDON, C.S.A., 1904,
REMINISCENCES OF THE CIVIL WAR, CHARLES
SCRIBNER'S SONS

A SOLEMN MEMORIAL

Edward Everett (1794–1865) was 69 years old when he was invited to speak at the dedication ceremony consecrating the National Cemetery at Gettysburg. Everett had been a Unitarian minister, president of Harvard University, a U.S. representative and senator, the 15th Massachusetts governor, the 20th U.S. Secretary of State, and a minister to Great Britain.

Local attorney David Wills (1831–1894) had been commissioned to purchase land for a cemetery dedicated to the burial of the Civil War fallen. Some eighty days after the battle, Wills invited Everett for a ceremony scheduled for October 23. However, Everett wanted more time to prepare. In deference to Everett's needs, Wills rescheduled the event for November 19, 1863.

Wills also wrote to the president to invite him to say a few words after Everett's oration. He wrote, "It is the desire that, after the Oration, you, as Chief Executive of the nation, formally set apart these grounds to their sacred use by a few appropriate remarks." Though Lincoln was the sitting president, Everett was recognized, at the time, as one of the leading orators in the nation.

> *"STANDING beneath this serene sky, overlooking these broad fields now reposing from the labors of the waning year, the mighty Alleghanies dimly towering before us, the graves of our brethren beneath our feet, it is with hesitation that I raise my poor voice to break the eloquent silence of God and Nature. But the duty to which you have called me must be performed; grant me, I pray you, your indulgence and your sympathy."*

> — EDWARD EVERETT, THE START OF HIS ORATION
> AT GETTYSBURG, NOVEMBER 19, 1863

LINCOLN AT GETTYSBURG

On November 18, 1863, President Abraham Lincoln took a short train excursion to southern Pennsylvania to attend the consecration ceremony scheduled for the following day. Accompanying him were:

- William Seward (1801–1872), his Secretary of State.
- John Usher (1816–1889), his Secretary of the Interior.
- Montgomery Blair (1813–1883), his Postmaster General.
- John Nicolay (1832–1901), his private secretary.
- John Hay (1838–1905), his assistant private secretary (officially a member of the interior department).
- William McDougall (1822–1905), a visiting dignitary from Canada.

On the way to Gettysburg, Lincoln told Hay that he was feeling weak. The following morning, he remarked to Nicolay that he felt dizzy. After returning to Washington, Lincoln was diagnosed with a slight case of smallpox.

Before leaving the house of David Wills on the morning of November 19, the president requested to meet John L. Burns, the only Gettysburg civilian wounded in the battle (see Chapter 11). The feisty senior citizen accompanied the president on his walk from the Wills house to the ceremony.

THE MEANING OF LINCOLN'S ADDRESS

Though Everett's two-hour oration was peppered with classical references to Ancient Greece, Plato, and empires long departed, Lincoln's short speech was based on references to the Bible and Christianity. Even the opening line—"Four score and seven years ago..."—has been tied to one of the Psalms.

Though we have numerous eyewitness sources for the wording of Lincoln's Gettysburg Address, the text differs in nearly all of them. While some phrases have been viewed by a few critics as controversial, like "under God," the fact that three of the reporters in attendance at the event telegraphed the entire text to include those words provides for us a strong argument that Lincoln did, indeed, invoke a reference to the Almighty.

The following is the most widely-accepted version of Lincoln's words at Gettysburg:

"Four score and seven years ago our fathers brought forth on this continent, a new nation, conceived in Liberty, and dedicated to the proposition that all men are created equal.

"Now we are engaged in a great civil war, testing whether that nation, or any nation so conceived and so dedicated, can long endure. We are met on a great battlefield of that war. We have come to dedicate a portion of that field, as a final resting place for those who here gave their lives that that nation might live. It is altogether fitting and proper that we should do this.

"But, in a larger sense, we can not dedicate—we can not consecrate—we can not hallow—this ground. The brave men, living and dead, who struggled here, have consecrated it, far above our poor power to add or detract. The world will little note, nor long remember what we say here, but it can never forget what they did here. It is for us the living, rather, to be dedicated here to the unfinished work which they who fought here have thus far so nobly advanced. It is rather for us to be here dedicated to the great task remaining before us—that from these honored dead we take increased devotion to that cause for which they gave the last full measure of devotion—that we here highly resolve that these dead shall not have died in vain—that this nation, under God, shall have a new birth of freedom—and that government of the people, by the people, for the people, shall not perish from the earth."

"Four score and seven years ago" refers to the signing of the *Declaration of Independence* in 1776, a little over eighty-seven years earlier. The remainder of the first sentence talks about the English and other immigrants seeking greater liberty and opportunity.

The next paragraph reveals Lincoln's concern that these United States can survive the war which had left the nation so divided.

The final paragraph starts with the humble reality that their efforts to consecrate the new cemetery fall short of the sanctification achieved by those who spilled their blood at Gettysburg. Instead, the living need to dedicate their lives to finishing the work of building and perfecting the country so nobly started by the Founding Fathers who had risked their lives nearly a century earlier and by the dead who had given their lives for the same purpose.

Some of America's most sacred phrases come from Lincoln's short speech; perhaps the two most memorable are "Four score and seven years ago" and "that government of the people, by the people, for the people, shall not perish from the earth."

TEST YOUR UNDERSTANDING

Answers to the following questions may be found in the Appendix.

1. Who was asked to give an oration at the Gettysburg dedication ceremony?

- *General Meade, commander of the Army of the Potomac.*
- *Edward Everett.*
- *William Wadsworth Longfellow.*
- *President Abraham Lincoln.*

2. Who traveled with Lincoln on the train to Gettysburg?

- *William Seward, his Secretary of State.*
- *Richard Montgomery, a minister of the British Parliament.*
- *General Meade.*
- *All of the above.*

3. At whose house did Lincoln spend the night before the ceremony?

- *John L. Burns.*
- *David Wills.*
- *John Hay.*
- *Lydia Leister.*

4. What illness did Lincoln suffer before and after the Gettysburg Address?

- *Chickenpox.*
- *Influenza.*
- *A nasty cold.*
- *A slight case of smallpox.*

5. To what does "four score and seven years ago" refer?

- *The end of the Revolutionary War.*
- *The ratification of the current Constitution.*
- *The implementation of the first central bank.*
- *The signing of the Declaration of Independence.*

SUMMING IT UP

Abraham Lincoln, as commander-in-chief of the military conducting the war, was invited to say a few words, but the main event was the superstar of oratory Edward Everett.

Millions of American schoolchildren have memorized Lincoln's few words, but the words of Everett have yet to be entirely known.

The Battle of Gettysburg is a solemn part of American heritage, but some myths must be dispelled.

13

GETTYSBURG LEGACY: MYTHS VS. FACTS

"Madam, don't bring up your sons to detest the United States government. Recollect that we form one country now. Abandon all these local animosities, and make your sons Americans."

— ROBERT E. LEE, QUOTED IN *THE LIFE AND CAMPAIGNS OF GENERAL LEE*, BY EDWARD LEE CHILDE (1875)

Myths of all kinds can spring into life after an important event. Some can be based on fact, others on conjecture, and still others entirely on imagination. Some myths might be planted lies to cover for some hidden truth. Quite often, the authenticity of a myth will remain unknown because there is too little information one way or the other.

The following myths are presented with the few facts we know about them. This is, by no means, an exhaustive list or a complete body of facts. Still, it does give the student of history a glimpse into the spin created by some who view history a bit too creatively or into the possibilities for which evidence no longer exists.

One thing is certain: Gettysburg was one of the bloodiest battles in history and certainly the most destructive of the Civil War.

. . .

Myth: Gettysburg Ghost Stories Are True

For over a century, only one known ghost story was circulated concerning the Gettysburg battle. It involved a place called Iverson's Pits close to Oak Hill. Then, in the 1990s, when interests in New Age philosophy and the Occult had been broadly popular for a generation, some started making money off the idea that Civil War ghosts were haunting the battlefield and the homes in the area.

Myth: Pickett's Charge Was the Greatest Attack of the Civil War in Terms of Size and Consequences

We have to view such judgments with a moderate amount of humility. American Battlefield Trust, for instance, suggests that the events at Chickamauga, Gaines' Mill, and Petersburg were more consequential than Pickett's Charge. We cannot merely weigh the number of deaths involved but also need to consider the effects of a battle on what followed—the advantages gained by one side or the other and their succeeding actions.

Myth: The Fighting on July 1 Was Not a Large Conflict

Though both Union and Confederate troops were still moving toward Gettysburg on the second day, the first day involved some incredible numbers:

- Almost 50,000 men were involved, and
- Roughly 16,000 were casualties (captured, missing, wounded, or killed).

Myth: Gettysburg Was Fought Because of Shoes

Why did General Heth send General Pettigrew to Gettysburg when their commander, General Lee, had expressly forbidden them from engaging with the enemy until all of them were gathered at Cashtown?

The topic of shoes was missing from the discussions of the motivation behind the Battle of Gettysburg until 1877. At that time, Henry Heth

wrote, "Hearing that a supply of shoes was to be obtained in Gettysburg, eight miles distant from Cashtown, and greatly needing shoes for
my men, I directed General Pettigrew to go to Gettysburg and get these
supplies."

One thing is sure: there were no shoe factories in the small town in
1863.

Could General Heth have been misinformed? Could he have lied to
protect his reputation after the fact? Or could there have been a supply
passing through on its way to the frontier? We do not know.

Myth: Confederate Private John Wesley Culp Had Been Killed on His Uncle's Property

Several myths swirl around this young lad—John Wesley Culp
(1839–1863), from Gettysburg, Pennsylvania—a man who had joined
the Confederacy and had earned the ire of many of the townsfolk. We do
not know if Henry Culp was his uncle, but the elder Culp owned Culp's
Hill and adjacent property. Some data suggests that the young Confederate private died on land east of Culp's Hill, not close to the elder
Culp's property. As of this writing, the story remains unverified and
thus a myth.

Another myth involving young John Wesley includes another
former citizen of Gettysburg—prisoner of war Johnston H. Skelly of the
Union Arm's 87th Pennsylvania regiment. This story tells us that young
Private Culp had attempted to deliver a message to Jennie Wade, Skelly's fiancée. As we learned in Chapter 11, Ms. Wade was the only civilian
fatality during the battle.

Myth: While riding the Train to Gettysburg, President Lincoln Finished His Speech on the Back of an Envelope

This widely circulated myth may have little basis in fact. Lincoln
traveled with three cabinet members plus a visiting dignitary from
Canada. This alone does not preclude any basis in fact. Still, when we
add the detail that the Library of Congress has the original copy of
Lincoln's Address on the stationery he used, the myth loses some cred-
ibility.

Did Lincoln jot some notes to himself on the back of an envelope? We do not know. While Lincoln worked on his speech before leaving the capital, he also worked on it on the evening before the event while he stayed at the home of Gettysburg attorney David Wills.

There is evidence that Lincoln may have ad-libbed as he spoke. His speech, as recorded by others in attendance, did not match the copies he had. And we know that he was not feeling well the evening before and for many days after the event; this may have affected his words on that day.

14

CONCLUSION: GETTYSBURG IN CONTEXT

The world might have been entirely different if General Lee had won the battle at Gettysburg. If he had gone on to terrorize the North, not only at Philadelphia but possibly even farther into enemy territory, perhaps Lincoln would have been forced to sue for peace. He may have been forced to resign.

General Robert E. Lee took an immense gamble by invading Pennsylvania. He was desperate to shorten the war, but the reluctance of first General Hooker and then General Meade acted to prolong the war for nearly two more years.

Gettysburg was a direct result of Lee's concerns over a war that should never have been fought. Those worries were a direct result of President Jefferson Davis's brash, unprovoked attack at Fort Sumter two years earlier.

The results at Gettysburg hinged on several mistakes, instances of disobedience, and command failures. Both armies were far from perfect on the battlefield. The haze of war thrived during those three hot summer days in 1863. And though many in the South pretended that Lee had accomplished great things, the iron resolve of the Confederacy had reached its limits and faltered.

Though General Ulysses Grant did not play a part at Gettysburg, he would later meet General Lee at Appomattox to accept the Southern general's surrender, effectively ending the Civil War. General Grant would go on to become the 18th president of these United States, promising his Southern brethren a good measure of compassion toward healing the nation.

Lee would return to his native state of Virginia, remarking to a friend, "I cannot desert my native state in the hour of her adversity.... I must abide her fortune, and share her fate." Lee became president of Washington College in Lexington, Virginia and served until his death of a stroke in 1870.

BIBLIOGRAPHY

Adelman, Garry E.; and Smith, Timothy H. (1997). *Devil's Den: A History and Guide.* Thomas Publications.

American Battlefield Trust. (ND). "Gettysburg." Retrieved on July 20, 2023 from https://www.battlefields.org/learn/civil-war/battles/gettysburg

American Battlefield Trust. (ND). "Robert E. Lee." Retrieved on July 23, 2023 from https://www.battlefields.org/learn/biographies/robert-e-lee

American Battlefield Trust. (ND). "Robert E. Lee's Decision to Invade the North in September 1862." Retrieved on July 20, 2023 from https://www.battlefields.org/learn/articles/robert-e-lees-decision-invade-north-september-1862

American Philosophical Society. (ND). "American Philosophical Society Member History (1878)." Retrieved July 19, 2023 from https://search.amphilsoc.org/memhist/search?year=1878;smode=advanced;startDoc=21

Belcher, D.W. (2018). *The Union Cavalry and the Chickamauga Campaign.* McFarland, Incorporated, Publishers.

Blount, Jr., Roy. (July 2003). "Making Sense of Robert E. Lee." *Smithsonian Magazine.* Retrieved on July 23, 2023 from https://www.smithsonianmag.com/history/making-sense-of-robert-e-lee-85017563/

Brands, H.W. (2012). *The Man Who Saved the Union: Ulysses S. Grant in War and Peace.* Anchor, an imprint of Penguin Random House.

Brandy Station Foundation. (1996). "The Battle of Brandy Station." Retrieved on September 28, 2007 from http://brandystationfoundation.com/pages/battle.htm

Burns, MD, Stanley B. (ND). "Civil War Medical Practice." Retrieved on September 7, 2023 from https://www.pbs.org/mercy-street/uncover-history/behind-lens/civil-war-medical-practice/

Carpenter, Layne. (April 4, 2018). "Fighting for their Lives: Medical Practices During the American Civil War — An Online Exhibit." Retrieved on September 7, 2023 from https://hsl.ecu.edu/2018/04/04/fighting-for-their-lives-medical-practices-during-the-american-civil-war-an-online-exhibit/

Chamberlin, Thomas. (1895). *History of the One Hundred and Fiftieth Regiment, Pennsylvania Volunteers, Second Regiment, Bucktail Brigade.* J.B. Lippincott Company.

Chapman, John Jay. (1921). *William Lloyd Garrison.* Atlantic Monthly Press.

Clark, Jesse. (February 17, 2023). "The Untold Truth Of Robert E. Lee." Retrieved on July 24, 2023 from https://www.grunge.com/223599/the-untold-truth-of-robert-e-lee/

Coddington, Edwin B. (1968). *The Gettysburg Campaign; a study in command.* Scribner's.

Crenshaw, Ollinger. (October 1, 1941). "The Knights of the Golden Circle: The Career of George Bickley." *The American Historical Review,* Vol. 47, No. 1. Oxford University Press.

Dixon, Ina. (October 29, 2013). "Civil War Medicine: Modern Medicine's Civil War Legacy." Retrieved on October 10, 2023 from https://www.battlefields.org/learn/articles/civil-war-medicine

Donald, David Herbert. (1995). *Lincoln.* Simon & Schuster.

Dowdey, Clifford. (1958, 2011). *Lee and His Men at Gettysburg: The Death of a Nation.* Skyhorse Publishing.

Dowdey, Clifford; and Manarin, Louis H., eds. (1961). *The Wartime Papers of R. E. Lee.* Little, Brown and Company.

Early, Jubal A. (1912). *Lieutenant General Jubal Anderson Early C.S.A.: Autobiographical Sketch and Narrative of the War Between the States. With Notes by R.H. Early.* J.B. Lippincott Company.

Eicher, David J. (2001). *The Longest Night: A Military History of the Civil War.* Simon & Schuster.

Emerson, Ralph Waldo. (1837). "The American Scholar," excerpts. Retrieved on July 17, 2023 from https://history.hanover.edu/courses/excerpts/111emerson-2.html

Encyclopaedia Britannica. (June 24, 2023). "Battle of Gettysburg: American Civil War [1863]." Retrieved on July 24, 2023 from https://www.britannica.com/event/Battle-of-Gettysburg

Gallagher, Gary W. (2001). *Lee and His Army in Confederate History.* University of North Carolina Press.

Gilje, Paul A. (1980). "The Baltimore Riots of 1812 and the Breakdown of the Anglo-American Mob Tradition." *Journal of Social History.* 13 (4): 547–564.

Gobat, Michel. (2018). *Empire by Invitation: William Walker and Manifest Destiny in Central America.* Harvard University Press.

Goellnitz, Jenny. (ND). "Civil War Medicine: An Overview of Medicine." Retrieved on September 7, 2023 from https://ehistory.osu.edu/exhibitions/cwsurgeon/cwsurgeon/introduction

Gordon, General John B. (1904). *Reminiscences of the Civil War.* Charles Scribner's Sons.

Gottfried, Bradley M. (2007). *The Maps of Gettysburg: An Atlas of the Gettysburg Campaign, June 3 – June 13, 1863.* Savas Beatie.

History.com (October 29, 2009). "Battle of Gettysburg." Retrieved on July 18, 2023 from https://www.history.com/topics/american-civil-war/battle-of-gettysburg

History.com. (October 29, 2009). "Robert E. Lee." Retrieved on July 23, 2023 from https://www.history.com/topics/american-civil-war/robert-e-lee

Jackson, Holly. (2019). *American radicals : how nineteenth-century protest shaped the nation.* Crown.

Kagan, Neil; and Hyslop, Stephen. (June 29, 2023). "Gettysburg was no ordinary battle." *National Geographic Magazine.* Retrieved on July 23, 2023 from https://www.nationalgeographic.com/premium/article/battle-of-gettysburg-day-maps

Longstreet, James. (1896). *From Manassas to Appomattox: Memoirs of the Civil War in America.* J.B. Lippincott Company.

MacKowski, Chris; and White, Kristopher D. (July 6, 2010). "Richard Ewell at Gettysburg." Retrieved on August 24, 2023 from https://www.historynet.com/richard-ewell-at-gettysburg/

McManus, John F. (December 13, 2018). "Overview of America," video. John Birch Society. Retrieved on July 14, 2023 from https://youtube.com/watch?v=tIl57cchRqs

McPherson, James M. (1988). *Battle Cry of Freedom: The Civil War Era. Oxford History of the United States.* Oxford University Press.

National Park Service. (ND). "Lydia Leister Farm: Meade's Headquarters." Retrieved on
September 5, 2023 from https://www.nps.gov/places/lydia-leister-farm.htm

National Park Service. (ND). "Robert E. Lee and Slavery." Retrieved on July 23, 2023 from
https://www.nps.gov/arho/learn/historyculture/robert-e-lee-and-slavery.htm

NetState.com. (ND). "Illinois State Motto." Retrieved on August 10, 2023 from
https://netstate.com/states/mottoes/il_motto.htm

New England Historical Society. (ND). "Boston Gentlemen Riot for Slavery." Archive
dated December 29, 2019. Retrieved July 19, 2023.

Ober, Josiah. (2008). *Democracy and knowledge: innovation and learning in classical Athens.*
Princeton University Press.

PBS. (ND.) "The Life of Robert E. Lee." *American Experience.* Retrieved on July 23, 2023
from https://www.pbs.org/wgbh/americanexperience/features/lee-timeline/

Pfanz, Harry W. (1987). *Gettysburg – The Second Day.* University of North Carolina Press.

Reilly, MD, Robert F. (April 29, 2016). "Medical and surgical care during the American
Civil War, 1861–1865." *Baylor University Medical Center Proceedings.*

Roshon, Jessica. (April 5, 2021). "Surviving, Persevering, and Profiteering: The Story of
Lydia Leister at Gettysburg," *The Gettysburg Compiler.* Retrieved on October 12, 2023
from https://gettysburgcompiler.org/2021/04/05/surviving-persevering-and-profi-
teering-the-story-of-lydia-leister-at-gettysburg/

Sartwell, Crispin. (January 1, 2018). "Anarchism and Nineteenth-Century American
Political Thought," chapter 16, *Brill's Companion to Anarchism and Philosophy.* Brill.

Sayers, Alethea D. (ND). "Introduction To Civil War Cavalry." Retrieved on September
28, 2023 from https://ehistory.osu.edu/exhibitions/Regimental/cavalry

Schiller, Laurence D. (ND). "The Evolution of Union Cavalry 1861–1865. Retrieved on
September 7, 2023 from https://www.essentialcivilwarcurriculum.com/the-evolu-
tion-of-union-cavalry-1861-1865.html

Sears, Stephen W. (2004). *Gettysburg,* Illus. ed. Mariner Books.

Sowell, Thomas. (2007). *A Conflict of Visions: Ideological Origins of Political Struggles,*
Revised ed. Basic Books.

Sowell, Thomas. (2009). *Black Rednecks and White Liberals.* Encounter Books.

Sowell, Thomas. (2012). *Intellectuals and Society: Revised and Expanded Edition.* Basic
Books.

Spooner, Lysander. (1845). "Works," including, "The Unconstitutionality of Slavery."
Retrieved on July 19, 2023 from https://lysanderspooner.org/works

Starr, S.Z. (2007). *The Union Cavalry in the Civil War: From Fort Sumter to Gettysburg,
1861–1863.* LSU Press.

Stone Sentinels. (ND). "The Leister farm – General Meade's Headquarters." Retrieved on
September 5, 2023 from https://gettysburg.stonesentinels.com/battlefield-
farms/leister-farm-meades-headquarters/

Sunrise Sunset. (ND). "July 2021 - Gettysburg, Pennsylvania - Sunrise and sunset calen-
dar." Retrieved on September 21, 2023 from https://sunrise-sunset.org/us/gettys-
burg-pa/2021/7

Thompson, Arthur R. (2018). "Myths vs. Facts" video series. Retrieved on July 17, 2023
from https://jbs.org/video/myths-vs-facts/

Thompson, Arthur R. (2018). *To The Victor Go The Myths & Monuments: The History of the*

First 100 Years of the War Against God and the Constitution, 1776 - 1876, and Its Modern Impact. American Opinion Foundation Publishing.

Thorsby, Gordon. (January 11, 2022). "Roder's Lone Gun at Gettysburg." Retrieved on August 24, 2023 from https://www.gordonthorsbycivilwarnotes.com/post/roder-s-lone-gun-at-gettysburg

Trefousse, Hans L. (1982). *Carl Schurz: A Biography,* University of Tennessee Press.

Victor, Orville J. (1863). *History of American Conspiracies: A Record of Treason, Insurrection, Rebellion, &c. in the United States of America from 1760 to 1860.* James D. Torrey, Publisher.

White, Jr., Ronald C. (2010). *A. Lincoln: A Biography.* Random House.

Wisconsin Historical Society. (ND). "Schurz, Carl (1829-1906)." Archive dated October 30, 2013. Retrieved July 19, 2023 from https://web.archive.org/web/20131030024230/http://wisconsinhistory.org/dictionary/index.asp?action=view&term_id=2626&search_term=schurz

Woodworth, Steven E. (2003). *Beneath a Northern Sky: A Short History of the Gettysburg Campaign.* SR Books.

ABOUT THE AUTHOR

Eric Porterfield is a father, lawyer, professor, and author. He holds a B.A. in government and German from the University of Texas at Austin and law degrees from Baylor Law School (J.D., valedictorian) and Harvard Law School (LL.M.). He writes in the areas of law, legal education, government, and history.

OTHER BOOKS BY AUTHOR

The American Revolution: A Concise History from Colonial Rebellion to the War for Independence to the Constitution, (2023)

The American Revolution - Available on Amazon